COAL MINE

Haynes

OPERATIONS MANUAL

First published in November 2020

A catalogue record for this book is available from the British Library.

ISBN 978 1 78521 714 2

Library of Congress control no. 2020931180

Published by J H Haynes & Co Ltd.,
Sparkford, Yeovil, Somerset
BA22 7JJ, UK.
Tel: 01963 440635
Int. tel: +44 1963 440635
Website: www.haynes.com

Haynes North America Inc.,
859 Lawrence Drive, Newbury Park,
California 91320, USA.

Printed in Malaysia.

All images are credited within the captions.

Author's acknowledgements
I would like to extend my deepest thanks to the staff of Big Pit National Coal Museum Blaenafon, Torfaen, Wales (https://museum.wales/bigpit/), for their enthusiastic assistance during the research and writing of this book. Special mention goes to the Curator (Coal Mining Collections) Ceri Thompson, who was unfailingly generous with his time, guidance and expertise, and without whose assistance this book would have been a far greater struggle to produce. Further thanks go to Matthew Saunders, for his engineering advice, and to Martyn Jones, Adrian Hawkins, Andrew Williams, Peter Richings, Ian Hart, David Trapnel and many other individuals who gave me their insight, even as a global pandemic began to make its presence felt.

I would also like to thank Joanne Rippin of Haynes for her flexibility and support during the publishing process, and for making books like this possible.

Dedication
I dedicate this book to the memory of my grandfather Ernest Davison, a miner in South Yorkshire, hardened by decades of toil and sharp experiences, but still a kind, wise and loving family man.

COAL MINE

OPERATIONS MANUAL

CHRIS MCNAB

CONTENTS

INTRODUCTION

The story of British coal mining is approximately 300 years old in industrial terms. If we trace the story to its ultimate source, however, the narrative begins hundreds of millions of years ago, in a subtropical prehistoric landscape.

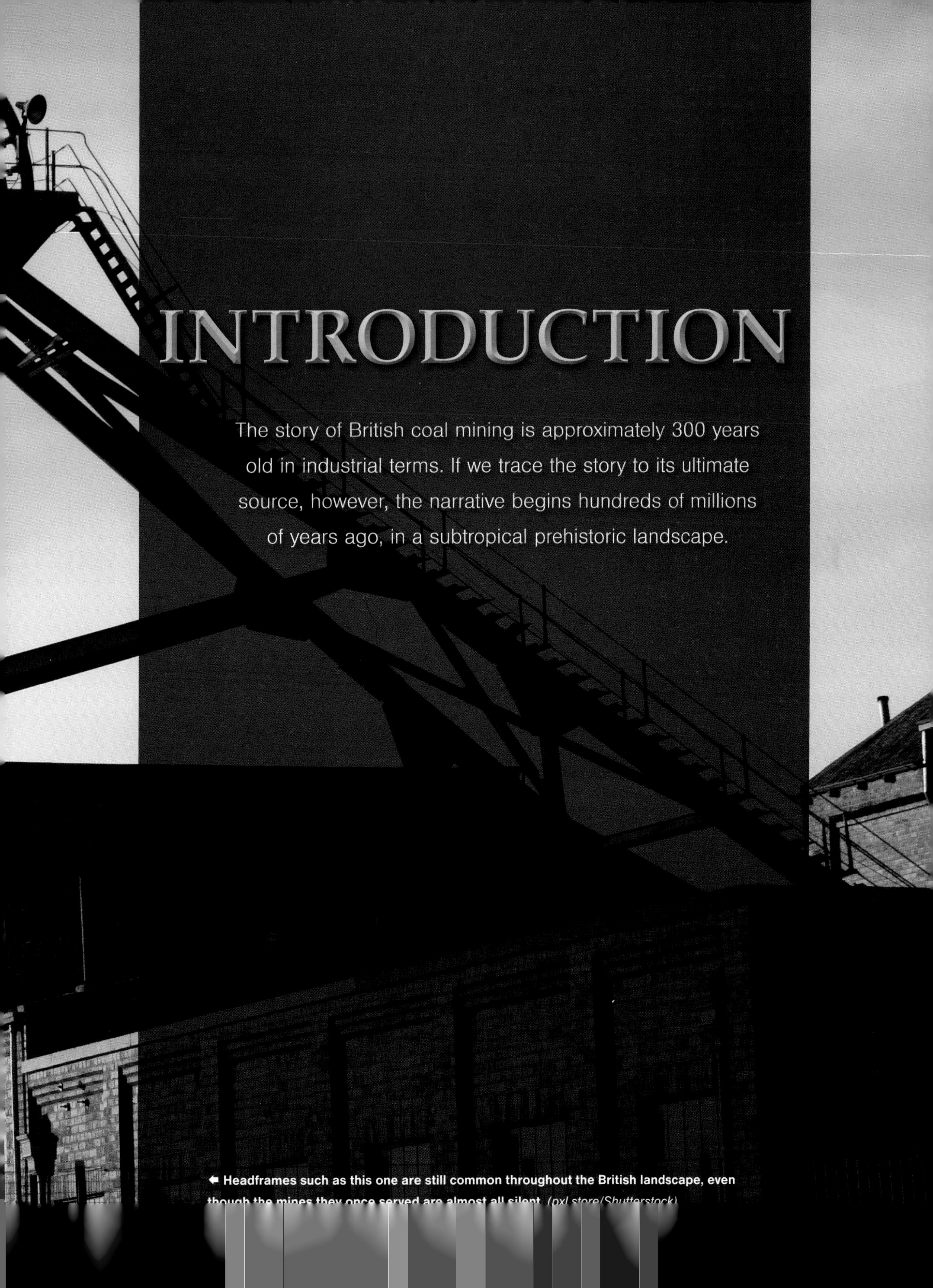

← **Headframes such as this one are still common throughout the British landscape, even though the mines they once served are almost all silent.** *(pxl.store/Shutterstock)*

One of the sources used in my research for this work is the optimistically titled *Coal: Technology for Britain's Future*. Published in 1976, it hit the bookshelves three years after the global oil crisis of 1973, triggered when the members of the Organization of Arab Petroleum Exporting Countries (OAPEC) enforced an oil embargo in the aftermath of the Arab–Israeli Yom Kippur War. In his foreword to the book, the then chairman of the National Coal Board (NCB), Sir Derek Ezra, not unreasonably looked to coal – abundant under the British landscape – as the stable future of the UK's energy supply. In short, 'The prospects for British coal are bright' (Ezra 1976: 7).

Looking back from the reflective vantage point of hindsight, Ezra's words have a defiant but rather tragic positivity about them. From the 1960s onwards, but accelerating in the 1980s, the British coal industry began what would be inexorable decline. It would be hard to see this coming, despite the fact that in many ways the coal industry actually began to contract from around 1921. In 1952, Britain's appetite for coal consumption peaked at 224 million imperial tons (228 million tonnes) per year, with 95 per cent of this volume produced in Britain via 1,334 deep underground mines and 92 surface mines. But from 1954, lengthening shadows fell across deep mining in particular, with production contracting by an average of 2.6 per cent every year from 1954 to 1983. Then, following the industry-wide impact of the national Miners' Strike in

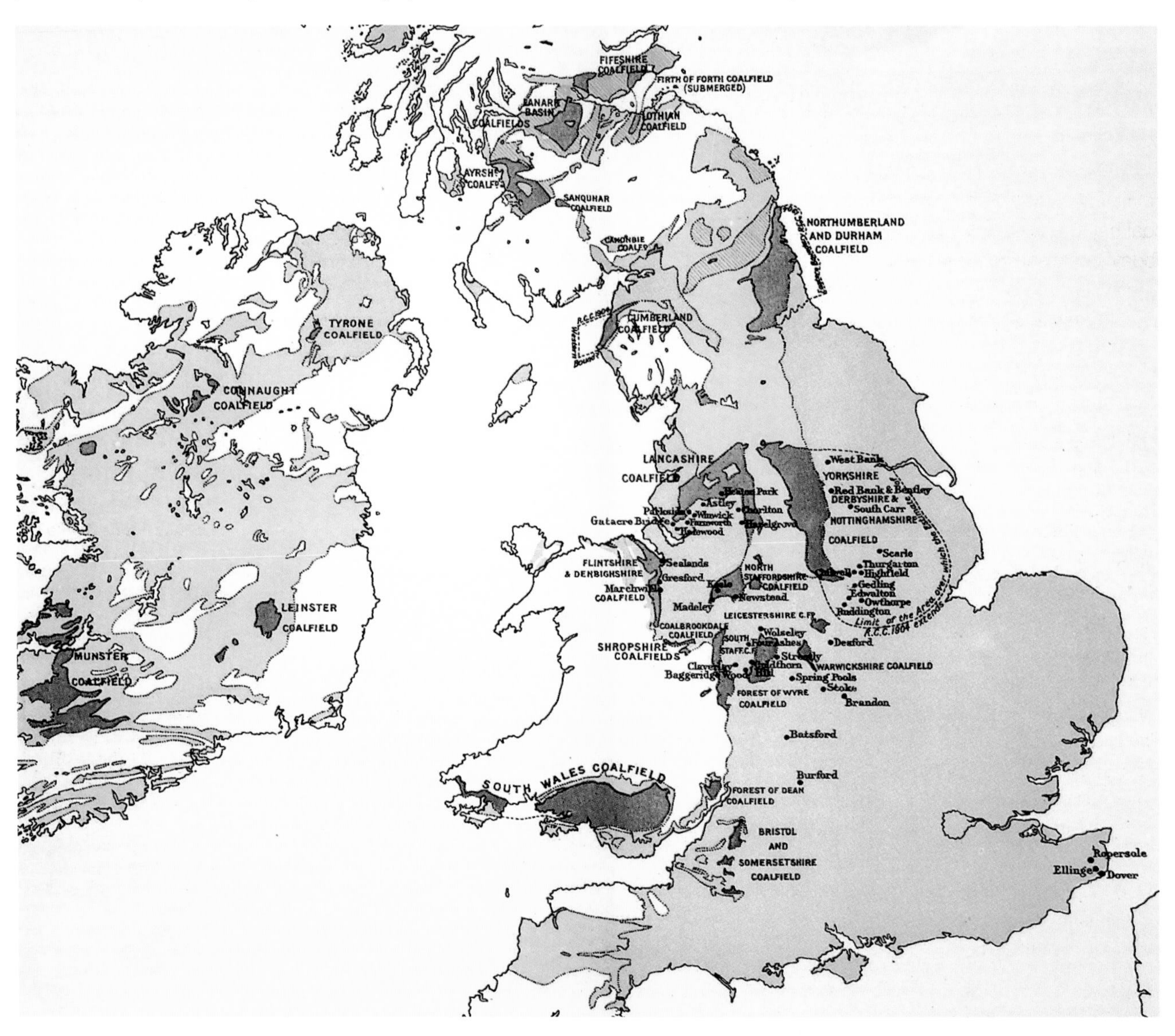

A section from a detailed map of British coalfields, in Boulton's *Practical Coal-Mining* (1908). The grey areas show the productive coal measures in the UK, illustrating how sparse and concentrated these were across the country. Blue areas show unproductive coal measures. *(Author/Boulton)*

1984–85, and subsequent colliery closures, production dropped by an average of 10 per cent every year; in 2013 it was down to 3.9 million tons (4 million tonnes). Notably, surface mining production actually increased by about 3 per cent each year from *c.* 1948 to 1991, with peak production in 1991 at 2.9 million tons (3 million tonnes), but declining thereafter by an average of 4.5 per cent each year from 1991 to 2005.

In the half-century from the 1960s, hundreds of coal mines closed, leaving behind both material and human consequences. In 2015, the last three of Britain's deep mines – Hatfield and Kellingley collieries in Yorkshire and Thoresby Colliery in Nottinghamshire – finally ceased production, truly the end of an era in Britain's industrial history.

Multiple reasons lay behind the decline of 'King Coal' in Britain. These included the rise of new energy sources, particularly oil and gas, the importation of cheaper coal from abroad, rising costs of production, political conflict and growing environmentalism, with coal being identified as a core global pollutant. Today, coal itself has little in the way of a positive brand image. Yet just a little historical digging quickly points to an undeniable reality: our modern age stands on the shoulder of this once-giant industry. For more than two centuries, coal powered the world. Without coal, there would have been no Industrial Revolution; without that, humanity would not have taken the giant leap into modernity that it did from the mid-18th century onwards. The infrastructure of this emerging new world hung upon coal – domestic heating, industrial and domestic power supply (including lighting from the production of coal gas), rail transportation, maritime and naval power, and manufacturing (including iron manufacture, using coke). It was also a vast source of employment for Britain's working classes; at peak employment in 1920, the industry gave labour to 1.19 million miners. Factor in all the miners' family members who relied upon their salaries, and we see the scale of the importance of mining not only to Britain's industrial health, but also to its cultural identity.

The aim of this book is primarily to give an accessible insight into the principles, practices and technologies of deep coal mining in the UK, concentrating mostly on the 18th to 20th centuries, but with some earlier historical context where relevant. Our narrative will run up to the demise of British underground coal mining in the 1990s, hence the persistent use of the past tense. Rather than moving through this history in a conventional chronology, however, we will break down the story into the logical areas of the mining process, looking at how each of these steps was performed throughout the industrial age.

Coal mining is a subject of endless technical breadth, so this book has had to perform much compression and summary. The focus is squarely upon mechanical and engineering aspects of mining, but the human story is always there powerfully in the background. My grandfather was a miner in South Yorkshire for his entire adult life, first going down the pit when he was 13 and only retiring from the industry

⬆ The still-functioning headgear at Big Pit National Coal Museum, in Blaenavon, Torfaen, Wales, formerly an active coal mine from 1880 to 1980. *(Author/ Big Pit NCM)*

⬇ A sample of bituminous coal, also referred to as 'steam coal'. This type of coal ranks between subbituminous coal and superior anthracite. *(James St John/ CC-BY-2.0)*

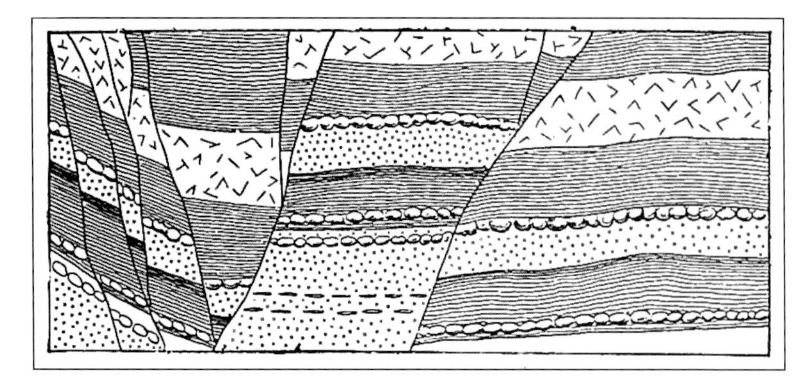

⬆ **A diagrammatic cross-section of rocks shows how faults can disrupt continuous horizontal beds of minerals.** *(Author/Boulton 1908)*

⬇ **The coal measures are replete with the fossilised remains of prehistoric creatures, with common types seen here.** *(Author/Boulton 1908)*

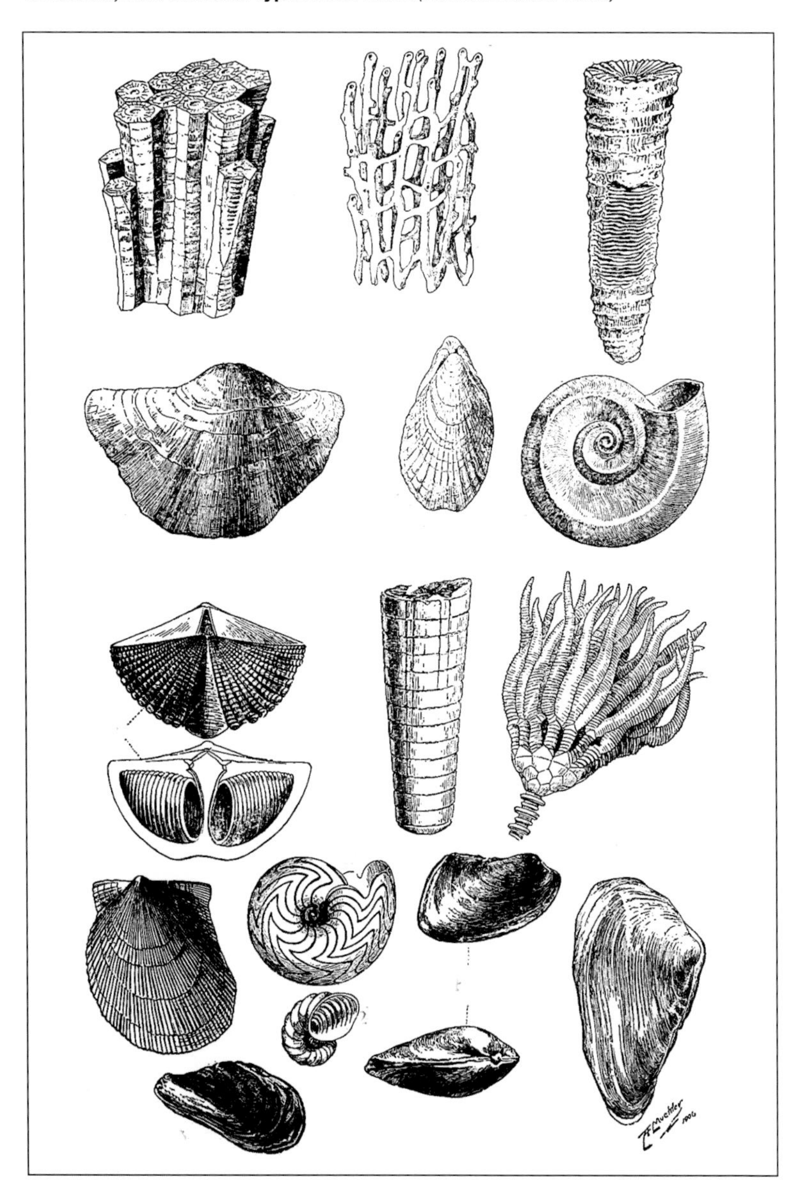

in his 60s. The stories he told me as a child embedded themselves deeply in my memory, and left me in no doubt that being a miner required resilience of both body and mind. He had little time for romanticism about the miner's life. But he, and millions of others like him, are part of a remarkable industrial generation, to which this book pays due tribute.

COAL FORMATION AND TYPES

Before we begin our exploration underground, we need an understanding of coal itself, and some of the challenges of getting to it. During the Carboniferous ('coal-bearing') period, between *c.* 360 and 280 million years ago, most of Britain was a swampy, subtropical landscape, with dense forest vegetation growing out of a silty, water-logged terrain. In a cyclical process over subsequent millions of years, forests rose up and died out, their organic materials (wood, roots, plant stems, leaf litter etc.) sinking down into the dense, mineral-laden mud. Land and sea levels also fluctuated under geological activity, inundating the landscape with seawater for long periods before retreating, the exposed land again becoming the site for new forest growth. Crucially for coal development, the silt and mud sealed the decaying vegetation from the atmosphere, preventing the bacterial actions of decay seen in surface-lying vegetation (which ultimately breaks plants down into humus), and trapping the carbohydrates and other elements that would eventually give the coal its unique energy-producing properties.

The buried vegetation compacted itself into successive layers of dense peaty material, sandwiched between rocky strata. Ultimately, aeons of time ended the cycle of peat formation, and much of Britain became dry land; the country had by this stage produced all the materials for coal formation that it ever would. Over the further millions of years that followed, thousands more metres of rock sat down heavily on top of the peat deposits, placing the ancient vegetation under enormous pressure and intense heat. The equation of pressure + heat + time, which gradually compressed the peat to about

1/20th of its original depth, converted it into the sedimentary rock we know as coal.

The strata in which the coal appears are termed Coal Measures; these include not only the coal layers or 'seams', but also the intervening layers of sandstone and shale dividing them. The typical depth of the Coal Measures in the UK is roughly 2,000–3,000ft (600–1,000m) thick. The actual workable coal seams, however, are typically 2–10ft (0.6–3m) thick, although they can be much deeper, occasionally tens of metres and, in some parts of the world, hundreds of metres thick. As a general rule, a layer of peat 120ft (36.5m) thick would compress down to a 6ft (1.8m) seam of coal.

Coal is not all of one type, industrially or materially. The level of pressure imposed upon the original peat, and the duration, has resulted in several different 'ranks' of coal. Arranging these in increasing orders of quality, they are:

- **Lignite** – Also known as 'brown coal', this soft, brown rock has a 60–70 per cent carbon content and a relatively low heat output. Lignite has more than 40 per cent content in 'volatile matter': referring to a mixture of gases, low-boiling-point organic compounds and tars. When burned, therefore, lignite also has the highest concentration of toxic emissions compared to other types of coal.
- **Subbituminous** – This coal is black in colour but dull in lustre, and has a higher heating value than lignite.
- **Bituminous** – This type of coal produces high heat and in appearance is smooth and shiny, with about 16–40 per cent volatile matter. (Subbituminous and bituminous coals are often used in electricity generation.)
- **Anthracite** – Occupying the highest rank of coal, anthracite is hard (hence it is often referred to as 'hard coal'), brittle and lustrous, with a high percentage of fixed carbon and a low percentage of volatile matter. Because of the energy density in this coal, as well as advantageous qualities such as moisture repulsion and relatively low levels of dangerous emissions, this coal was the ideal type for domestic and industrial applications.

⬆ Open-cast mining takes advantage of coal deposits situated at or close to the surface. Here we see a Russian 'brown coal' mine in operation. *(kemdim/ Shutterstock)*

⬅ Lignite coal or 'brown coal' is a low-quality fuel used almost exclusively in steam-electric electricity generation. *(Edal Anton Lefterov/CC BY-SA 3.0)*

Crucially for what would become the coal industry, seismic activity over endless horizons of time pushed coal seams towards the earth's surface, making them accessible to varying degrees and methods. At best for the miner, a seam presents the strata of coal horizontally and in unbroken fashion over a long distance, but geology is rarely so neat. More commonly, the seam will undulate up and down, the angle of rise and fall ranging from gentle inclines to almost vertical presentations. Sometimes a 'fault' – essentially a crack in the earth's crust – splits the coal seam, so that the seam abruptly stops against bare rock, continuing at a higher or lower level. Coal seams can perform all manner of other buckles, twists, turns and deviations. The volumes of other rock types also vary in and around the seam, further complicating the business of coal extraction.

The underground world of the coal mine is often a journey back into prehistory. Here we see an entire fossilised tree trunk, towering upwards in a Welsh seam. *(Author/Big Pit NCM)*

The geological factors related to the presentation of coal have meant that the accessibility and volume of coal is concentrated in particular regions of the UK, as 'coalfields'. Summarising broadly, the major coalfields were to be found in South Wales from Monmouthshire to Pembrokeshire, the Midlands (principally Staffordshire, Derbyshire and Nottinghamshire), north-west and north-east England, and central Scotland. Within each of these areas, Coal Measures are classically divided in Upper, Middle and Lower types. The actual geological composition of these strata is complex and variable, but each coalfield possessed its own characteristics and mining challenges. By way of illustration, here is a description of just one coalfield, from the multi-volume 1908 work *Practical Coal-Mining*:

> *'VI.* ***Northumberland and Durham.****—This coal-field extends through the eastern part of the counties of Northumberland and Durham for a length of about 60 miles, and for at least 3 miles under the sea. Of the 60 coal-seams or more known to exist, about 23 are without doubt capable of being worked.*
>
> *Along the western margin of the field the coals dip towards the east, but near the coast-line they are nearly level, and even in some cases rise towards the surface. [. . .] the Coal-Measures proper may be divided into: —*
>
> 1. Middle and Upper Coal-Measures, *2000 feet thick in Durham, containing all the important seams; nodules and bands of ironstone; sandstones suitable for building purposes; and the "grindstone post", from which the famous Newcastle grindstones are made.*
>
> 2. Lower Coal-Measures, *350 feet thick, of which 150 feet consist of ganister beds with two seams of coal. These rest upon the Millstone Grit, which has considerably thinned out in tracing it from the south.' (Boulton 1908: 56)*

The authors go on to explain in turn the other coalfields in the UK, each with its local set of idiosyncrasies. The South Wales coalfield, for example, 'is a basin with its longer axis east and west, the measures on the northern crop dipping gently to the south, while on the east and south their edges turn up sharply at a much steeper angle' (Ibid: 52). It was partly the unique nature of each coalfield, combined with local cultural traditions and conventions that gave mining in the UK such a distinct regional identity, rather than being a homogenous industry. What did unite every colliery and miner, however, was that they had to find or implement the most efficient methods of extracting this precious prehistoric material and getting it to the surface.

➡ **The dramatic but now disused headstocks of Clipstone Colliery in Nottinghamshire.** *(Dave Beavis/CC BY-SA 2.0)*

⬇ **The headgear of Hope Pit at Caphouse Colliery (Overton Colliery) in Overton, near Wakefield, West Yorkshire.** *(J3Mrs/CC BY-SA 3.0)*

MINE LAYOUT

In a coal mine, there was an almost organic relationship between all the working parts. Mine planning was in many ways akin to town planning, the engineers and managers attempting to achieve a seamless flow of activity from the coalface to the shaft top and across the surface workings.

⬅ **By the 20th century, it was law that coal mines had two points of entry/egress, most commonly reflected in the dual shaft arrangement seen here, each shaft with its own**

PRE-INDUSTRIAL CONTEXT

The history of coal mining in Britain is ancient or relatively modern, depending where you draw the historical line. There is evidence of minor coal use dating back to the Bronze Age (*c.* 3200–600 BC), including for cremation rites, and it was burned by the Romans for heating, metalworking and glass-making during their occupation of Britain from AD 43. (It is worth pointing out that by this time, China had a fully emergent coal industry; Europe was simply lagging behind in this regard.) Yet coal use was truly minor in comparison to that of wood and charcoal; coal in Roman Britain was unlikely to be mined in any sort of conventional sense, but simply gathered from exposed outcrops or as pieces washed up on beaches and coastlines (so-called 'sea coal').

Following the Roman exodus from Britain in the 5th century, coal largely drops out of the historical record in Britain until the 1200s. During this century, there are early records of monks in Scotland and northern England quarrying coal from outcrops, using it to power iron forges. Civilian use of coal also expanded a little, and by the 15th century coal mining had become a protean industry, dotted around the country and expressed through the simplest forms of drift mining and bell-pit workings (see below).

It was from the 16th century, however, that coal began to gain more of a presence in the country's hearths and furnaces. The 'Timber Crisis' – a critical lack of good timber, resulting from Britain's massive deforestation for firewood – drove up coal's value and gave it greater recognition by the royalty and nobility. Coal also came to have wider applications in society, not only in domestic heating (aided by the development of the household chimney in the Tudor period), but also in minor industries such as glass-making, brewing, baking, lime burning, salt evaporation and brass-founding (Preece 1981: 8). Mining sites proliferated, first with major concentrations in the north-east of England, but gradually spreading elsewhere. Some headline figures give an impression of the expansion of the coal industry. In 1660, the annual output of coal was some 6 million tons (6.09 million tonnes), but it climbed to 7 million tons (7.11 million tonnes) by 1730. One hundred and seventy years later, in 1900, the output was 200 million tons (203 million tonnes).

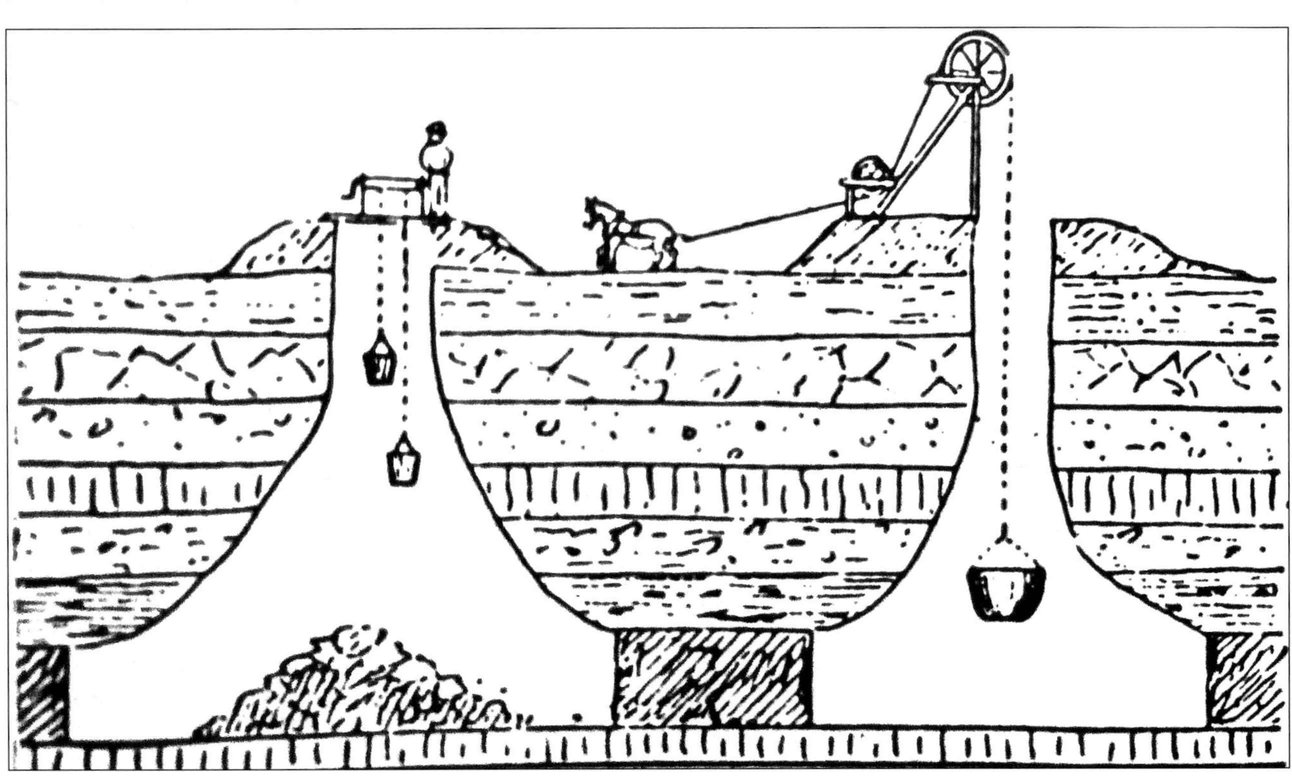

A diagram of early bell-pit mining, showing coal lifting by either hand-powered windlass or horse-drawn pulley. *(Author/Big Pit NCM)*

FINDING THE COAL

Before coal can be mined, it has to be found in the first place. For much of mining history until the 17th century, the primary survey method for discovering the presence of coal was visual – simply spotting the black lines of coal outcrops in the landscape, these often found following a natural feature such as a coastline or river valley. Applying some early geological and logical extrapolations to this information meant that locating simple coal mines could be more than mere guesswork. Finding productive seams, however, was another matter, pursued mainly by sinking exploratory shafts, a particularly time-consuming and expensive process. As the demand for coal expanded in the 16th and 17th centuries, it was clear that more advanced methods were required for finding the richest coal seams, especially those locked deep underground.

An important step forwards was in the development of boring technologies from the Early Modern period. With these, the mine prospector used cutting and drilling equipment to sink a narrow but deep borehole into the ground, from which geological samples were extracted. The samples in turn guided the decision about whether to establish a full-blown mine at the site. (It should be noted that boreholes have historically also had a range of other purposes in mining, including draining off or raising water, installing communication lines, guiding haulage ropes and directing steam or compressed-air pipes underground.)

Taking a slice through history at *c.* 1900, the most basic boring technology was a simple steel chisel, the non-cutting edge threaded to fit on to guide rods. The type of cutting edge varied according to the material challenge. For example, a V-shaped diamond bit was most suited to punching through hard rock. For working through softer materials, such as clay, the cutter might be changed to one of more cylindrical types, which not only cut but also channelled loose material up past the cutting edge. Periodically, the borehole would need to be cleared of debris; for this purpose a 'Sludger' was used, a tubular

⬆ Miners and engineers prepare boring rods to conduct a trial boring for coal. Proper configuration of rod joints was essential to avoid rod fractures and distortions. *(Geoff Charles/CC0)*

⬅ A 'sinking bucket' was using during the sinking of a mine shaft, the bucket used to lower and raise workers, tools and debris. *(Author/Big Pit NCM)*

instrument that was dropped down the hole and filled up with sludge, which could then be drawn to the surface.

The design of the boring rods was also critical. They were made of either iron or wood, about 1in (2.5cm) square and fastened together in threaded sections. Although iron rods were certainly stronger, wooden rods were lighter, and this mattered when dozens or even hundreds might be connected; the combined weight of those above risked shattering the rods below, or causing them to bend or spring. This solution was partly discovered through the use of sliding-joints, or 'free-falling cutters'. First invented in the 1830s, this mechanism essentially inserted a cylindrical sliding joint between the chisel and the lowest rod; the chisel body slid within the slot of the sliding joint on impact. The fact that the sliding joint was longer than the drop of the rods meant that the blow on the rock was delivered independently of the weight of the rods.

In the first two centuries of mine surveying, the physical action of boring was performed arduously by hand. The rods typically terminated at the top in either a single brace-head (operated by two men) or a double brace-head (operated by four men). *Practical Coal-Mining* notes that 'Two men with a single brace-head can put down a hole of about 40 to 50 feet; and with a double brace-head, allowing four men to work at it, a depth of about 80 feet may be reached' (Boulton 1908: 100). Yet the efficiency and depths of borehole surveying was dramatically increased by mechanisation. From the mid-19th century, steam-powered boring machines came into service, these ramping up the speed, power and consistency of percussive boring. Furthermore, in 1873 mechanised rotary boring was introduced, using a rotating diamond-encrusted drill bit. By the early 20th century, borers were allied to electrically powered systems, again with efficiency gains.

A mining surveyor takes levels to assess the extension of a deep roadway. *(Adwo/Shutterstock)*

It is impossible here to list all the varieties and mechanisms of mine survey boring systems. Yet we can note that drilling is still a primary means of coal exploration. There are two main methods: 1) core drilling, which uses a hollow drill bit that collects cored samples of the strata through which it passes; and 2) rotary drilling, which collects samples of the rock that have been pulverised by the passage of the solid drill bit. The advantage of rotary drilling is that it is faster and cheaper than core drilling; for this reason, it is not uncommon for rotary drilling to be used to reach the coal seam, when the engineers switch to core drilling to gain accurate samples.

In the final years of the British coal industry, mining engineers acquired a range of ultra-sophisticated tools to aid geophysical surveys, using advanced technologies to locate and map underground coal deposits with high levels of accuracy and analysis. These roughly divide into airborne methods (i.e. technologies mounted on an aircraft) and ground methods. The most common were (indeed remain so):

Airborne

- electromagnetic
- magnetic plus electromagnetic
- magnetic plus radiometric
- magnetic

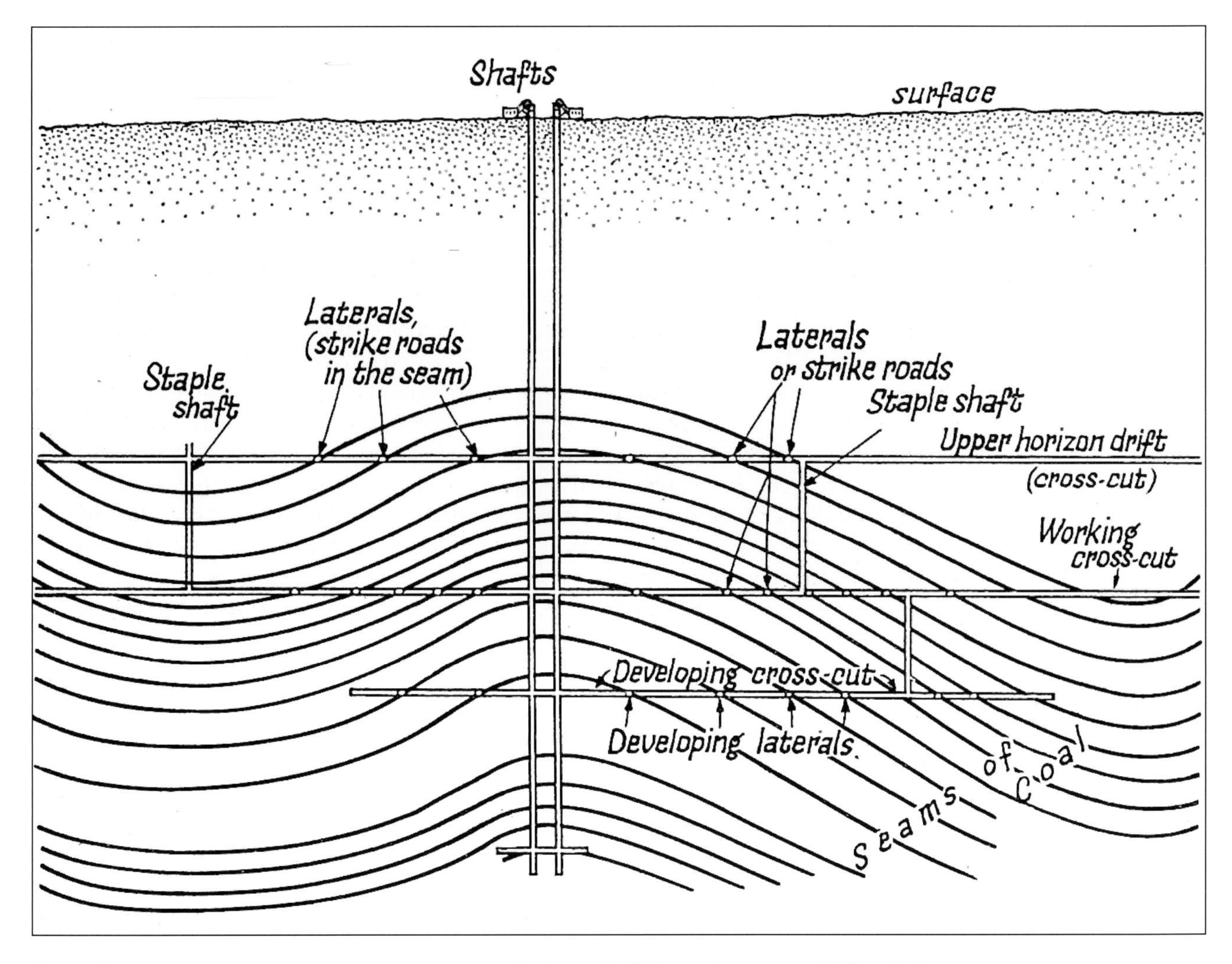

A graphical representation of horizon mining, with the level drifts (or horizons) cut out from the shafts through the densest seams. *(Author/Mason)*

Ground

- electrical
- magnetic
- electromagnetic
- radiometric
- gravimetric
- seismic refraction

To take one as an example, seismic refraction works by generating seismic waves, typically by percussive, explosive or projectile means, and directing them into the ground. Sensitive electronic devices then detect and analyse the wave refractions that bounce back off different strata of geological material below.

Looking back in history, however, the old days of percussive and rotary boring were far more analogue in nature. It was also a process that took time before confident data was available. For example, during the explorations in and around Shakespeare Colliery in Kent in the early 20th century, some 40 boreholes were drilled between 1905 and 1914, just to ascertain the extent of the coalfield.

BELL-PITS

The earliest form of underground mine was the bell-pit, although in many ways it has just as much in common with surface mining as deep mining. Bell-pits first appeared during the 15th century, as outcrop coal resources began to appear limited, and they were the main form of coal extraction until the 17th century.

A bell-pit consisted of a simple shaft dug straight down into the earth, reaching to a depth of anything between 32ft and 295ft (10m and 90m), the depth largely depending of the proximity of the coal seam and the ability of the ground to hold the shaft. The bottom of the shaft would be widened out around its circumference as the coal was

An 18th-century cross section of a mine, plus furnaces and related equipment below. The mine configuration, with its access shaft, supported roads and working faces, is becoming recognisably modern. *(Wellcome Collection Gallery/ CC BY 4.0)*

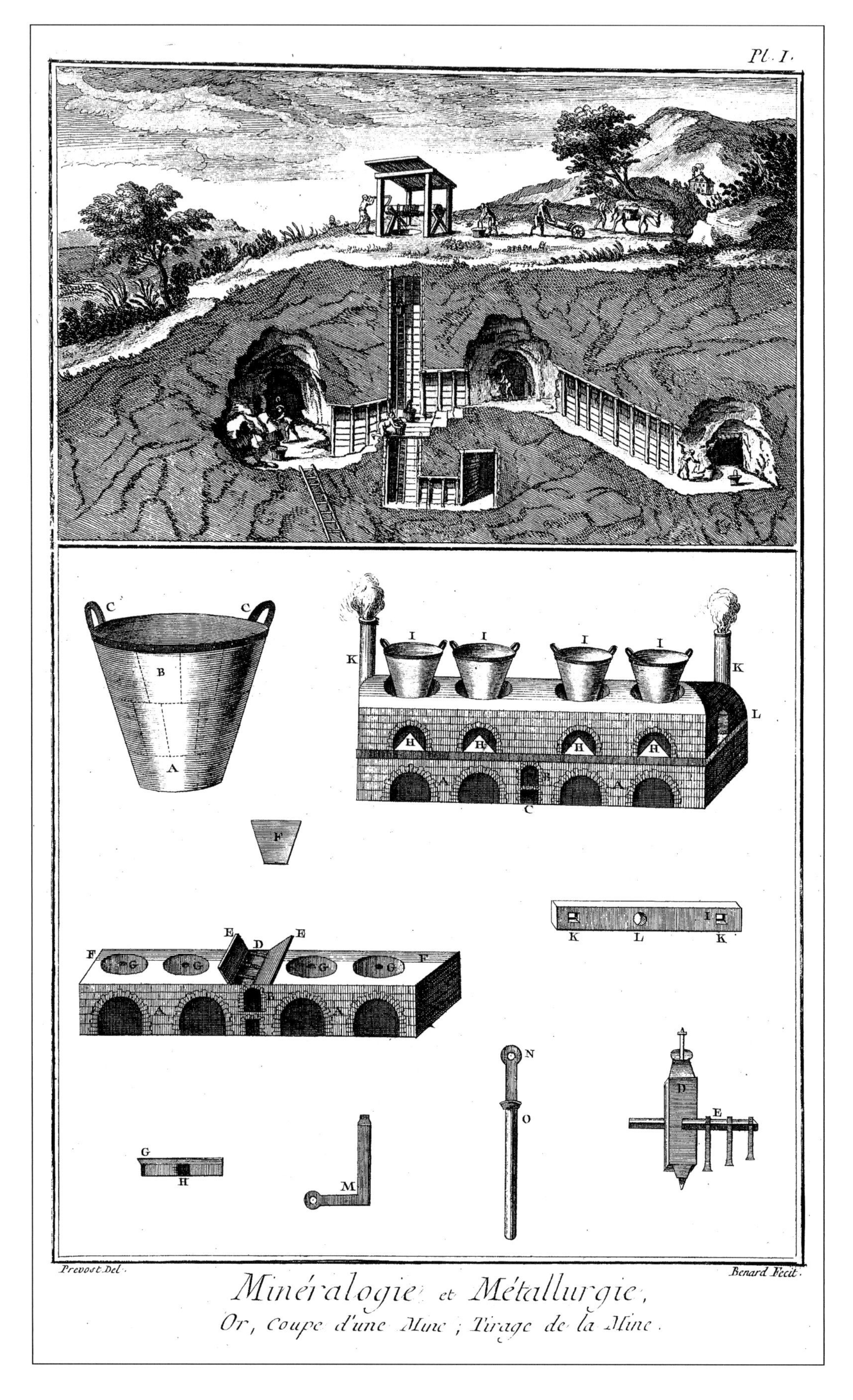

mined, eventually producing a shape that in profile looks somewhat like a wide-bottomed vase with a narrow neck. The extent to which the mined area flared out depended on how far the ground could be undercut without the roof collapsing; the bell-pit walls and roof were not lined or supported. Once the bell-pit was adjudged to be too risky to operate, the miners simply moved to an adjacent location and began digging a new bell-pit.

Hacked out from the earth, bell-pits were crude and dangerous affairs. Ascent and descent was either by ladder or via the rope system that was also used for raising and lowering the extracted coal in baskets. The rope systems used included simple two-man windlasses or horse-operated pulleys. Hazards of the bell-pits were the build-up of poisonous gases, flooding, roof or shaft collapse, and falls down the shafts.

DRIFT MINING

If geological conditions permitted, an alternative to the bell-pit was the drift mine. This underground mine was formed by driving an adit or a drift – a horizontal or near-horizontal tunnel – into a hillside, following the line of the coal seam. At first, the unsupported adits ran relatively shallow underground, but during the 17th century tunnels were often given roofing supports, enabling the drift mine (as it was called) to extend deeper and deeper into the landscape.

Although drift mining never reached the scale of production achieved by deep coal mines, it would nevertheless form a major part of coal-mining history in the UK and certainly abroad (drift mining was especially extensive in the USA). Adits came to be very long indeed, not only serving as passageways to the coalface but also as drainage channels to keep the drift mines free from flooding. (Ideally, adits should be constructed with a slight upward incline to allow the free drainage of water out of the mine.) The Milwr Tunnel mine drainage adit in North Wales, for example, extends a total of 16km (10 miles), reaching from the hamlet of Cadole near Loggerheads, Denbighshire, to Bagillt on the Dee Estuary.

The gated mouth of an old drift mine in Beamish, at The North of England Open Air Museum, in County Durham. *(Cory Doctorow/CC BY-SA 2.0)*

SINKING A SHAFT

As the 17th century progressed, the coal of overworked bell-pits and early drift mines became increasingly exhausted. As a consequence, the expanding mining industry began to look at vertical depth as the future. Mine shafts were consequently sunk deeper and deeper to access underground seams. The depth to which the shafts could be sunk, however, was limited by two fundamental problems in mining: drainage and ventilation. Crude manual techniques for managing these problems (see Chapter 5) meant that until the early 18th century, mine shafts tended to be restricted to a maximum depth of 350ft (105m) and a radius from the base of the shaft of not more than 600ft (180m). The invention of steam engines, however, led to far better solutions to the drainage and ventilation issues, and consequently mines started to go deeper and deeper into the earth.

There are a handful of important points to note before we delve deeper into shaft construction. First, from 1862, British deep mines were legally required to have two shafts located close to one another, to allow the circulation of air through the pit (see Chapter 5) and to facilitate rescue if one of the shafts were put out of action. Second, individual colliery shafts serve a variety of purposes. Shafts were used for raising coal, ventilation,

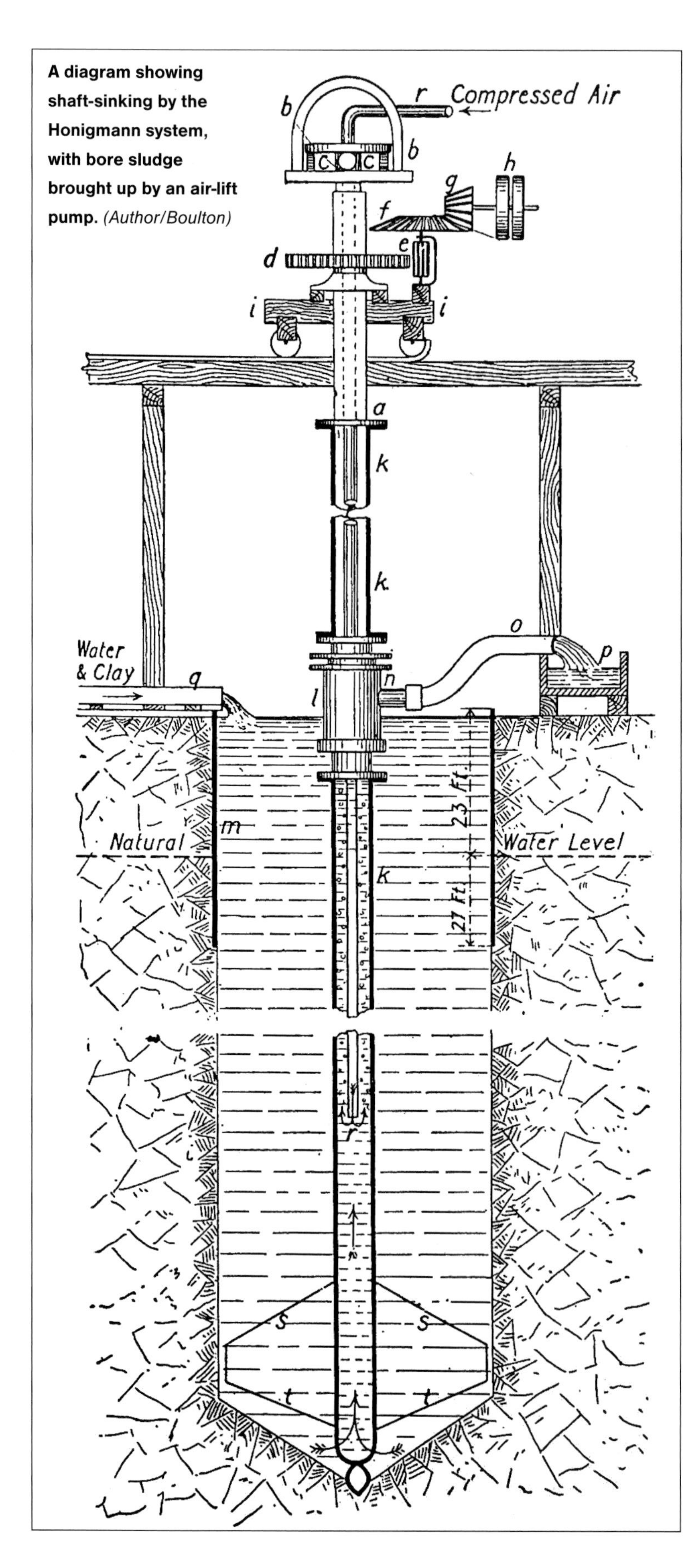

A diagram showing shaft-sinking by the Honigmann system, with bore sludge brought up by an air-lift pump. *(Author/Boulton)*

transporting men and materials, water pipes, compressed air pipes etc. – the shafts were split vertically if there was only one shaft, so that half acted as an intake and the other half for the used air – this was done by wooden shuttering or brattice.

Whatever the era, shaft sinking has always faced a common set of challenges, specifically:

- excavation of the rock
- supporting the sides of the shaft as it is cut
- removing the excavated material
- handling water drainage
- ensuring the shaft is functional for humans (lighting, ventilation etc.)

MAIN PARTS OF A MINE SHAFT

Headgear/headframe – Also 'winding tower' or 'pithead'. This is the visible part of the shaft system above ground, typically consisting of a motorised hoist mechanism, mounted on a framework, used for winding cages of men and material up and down the mine shaft.

Shaft collar – Also the 'bank' or 'deck'. The shaft collars consist of the uppermost portion of the shaft. It supports the weight of the headframe, acts as a protective barrier to prevent water and soil from entering the shaft, and provides the necessary structure for men and materials to enter and leave the shaft. A plenum space (a pathway for air circulation) – also referred to as a 'fan drift' – is incorporated into the upcast shaft collar to allow ventilation.

Shaft barrel – The shaft barrel is the part of the shaft that connects to the collar and which runs down for most of the length of the shaft underground.

Shaft station – This is the point where the shaft barrel meets the horizontal workings of the mine. It is where miners, minerals or other services enter or exit the shaft at the bottom, and thus includes loading systems.

SHAFT DIGGING

Early shafts were dug the hard way, with manual tools such as picks, shovels and rock-splitting wedges. Later, hammer-struck boring drills were used and, from 1749, we see the use of explosives for shaft cutting. The shafts produced were anywhere between 4½ft and 12ft (1.4m and 3.6m) in diameter, and were circular, square or oval in cross-section. The majority of shafts in British coal mining history have been circular, with square-profile shafts typically only used in ground that was dry and easily worked. Mine shafts might be sunk vertically – the most common type – or on an incline, depending on problems of access.

The challenge of shaft sinking was, as with all aspects of mining, improved by progressive mechanisation. During the 18th century, steam-powered hoists and pumps improved the pace of drainage and debris clearance, and in the 19th century came pneumatic rock drills. Electrically powered percussive drills entered service in the early 20th century. The subsequent advances of the 20th century utterly transformed the processes of shaft sinking – mechanical mucking and shaft excavation machines; hydraulic drills (replacing pneumatic ones); more efficient explosive types that were easier to handle; and shaft jumbos (large-scale drilling rigs used to cut multiple holes for explosive placement). The result was, over time, a pronounced increase in the speed of shaft sinking. For example, between 1600 and 1800, the speed of shaft digging was about 32–39ft (10–12m) per month, but from 1900–40 it increased to 98–130ft (30–40m) per month. Advancing

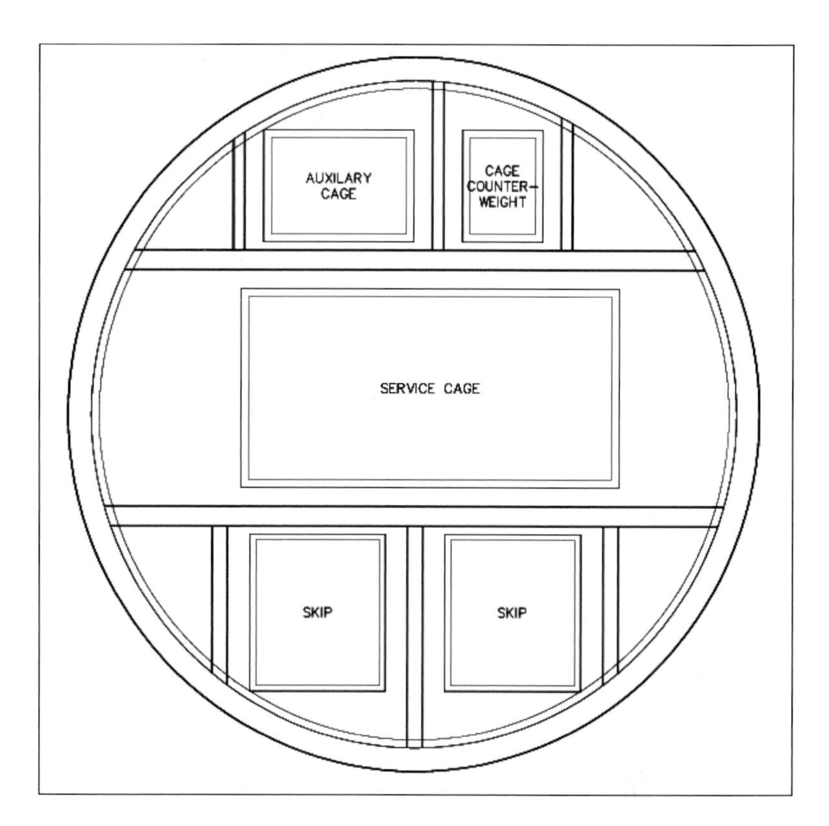

A typical mine shaft plan, showing how the shaft is vertically arranged into separate working spaces.
(Kelapstick/ CC BY-SA 3.0)

The pit bank at Big Pit National Coal Museum, Wales. The shaft below descends 300ft (90m).
(Author/Big Pit NCM)

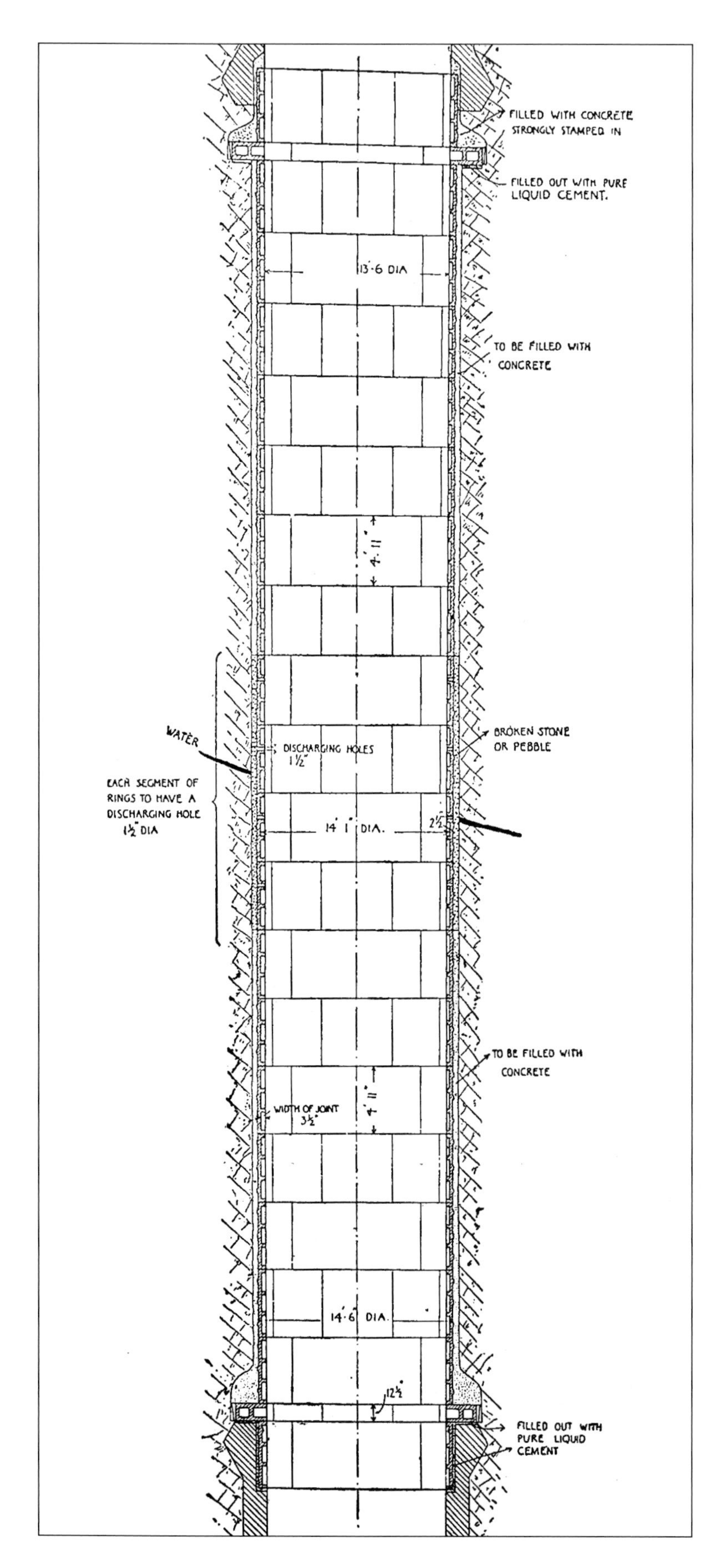

This diagram shows the German tubbing system applied to Llangennech Colliery in South Wales. *(Author/Boulton)*

from 1940 to 1970, however, the rate stepped up again to 295–328ft (90–100m) per month, a pace that has held to this day.

The complexity of shaft sinking depended much on geological conditions. For example, if the shaft was being cut through solid, non-frangible rock, or through strata that didn't produce a heavy influx of water, it might simply be cut to shape and left unlined, the exposed rock forming a natural wall. Such circumstances were relatively rare, however. Instead, the shaft sinkers might find themselves cutting through mixtures of hard rock, soft rock, loose soil, sand and other materials, all with copious amounts of water jetting into the shaft space as it was opened up. For this reason, most shafts were lined.

SHAFT LINING

The earliest form of lining was simply brickwork built up around the shaft wall, either mortared in place or stacked 'dry' (i.e. unmortared), so that the bricks could be repurposed at a later date. To counter loose earth or heavy water ingress, however, 'tubbing' was applied. This comprised, initially, of struts of timber arranged around the wall of the shaft in a manner similar to barrel construction, the wood wedged in place by wooden or metal rings. In 1841–43, however, the first use of cast-iron tubbing occurred, during the construction of Snibston Colliery in Leicestershire, and this method became standard well into the 20th century.

Brickwork could be constructed with a basic degree of water resistance, especially in the 'coffering' method of lining, which used concentric rings of bricks attached by a continuous binding of hydraulic cement. The cast-iron tubbing, however, came in sections that bolted together, tightening up to form a true watertight seal. During the 20th century, other materials were also used as shaft linings, including precast concrete sections and shotcrete (a concrete or mortar pneumatically projected at high velocity on to a surface via a

FREEZING THE SOIL

A special problem for shaft sinkers was how to cut into ground that was heavily sodden with water. An ingenious solution was hit upon by the German engineer F.H. Poetsch in 1883. *Practical Coal-Mining* describes the mechanism thus: 'It consists of putting down a number of tubes from 2 to 4 feet apart, so arranged as to form a ring several feet larger in diameter than the finished shaft; through these tubes a freezing-mixture is forced to circulate, which on its upward path extracts sufficient heat from the water-bearing strata, to freeze the water contained in them. By this means the running ground is converted into a solid mass in which sinking can be carried out in the ordinary way until the water-bearing strata are traversed. A curb is then laid in the impervious rock, and upon this a brick- or iron-lining is built up. When the latter is completed, the tubes are thawed out and drawn' (Boulton 1908: 219–20). The freezing mixture referred to was classically brine, and the same mixture is used today in modern ground-freezing technologies, although liquid nitrogen can be applied as a more expensive but efficient option.

An early 20th-century plan of a ground-freezing plant, designed to pump frozen brine into the soil structure. ***(Author/Boulton)***

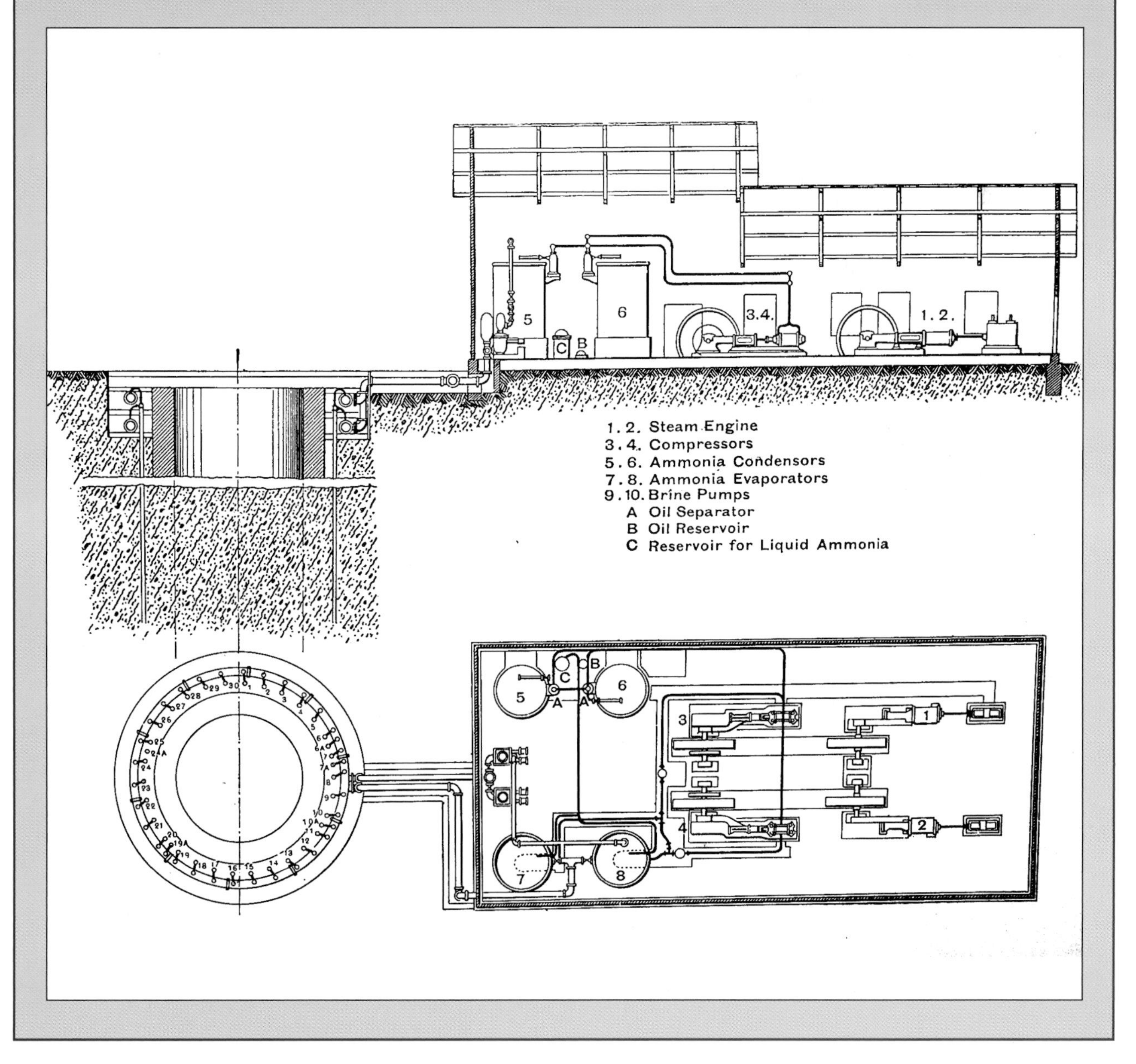

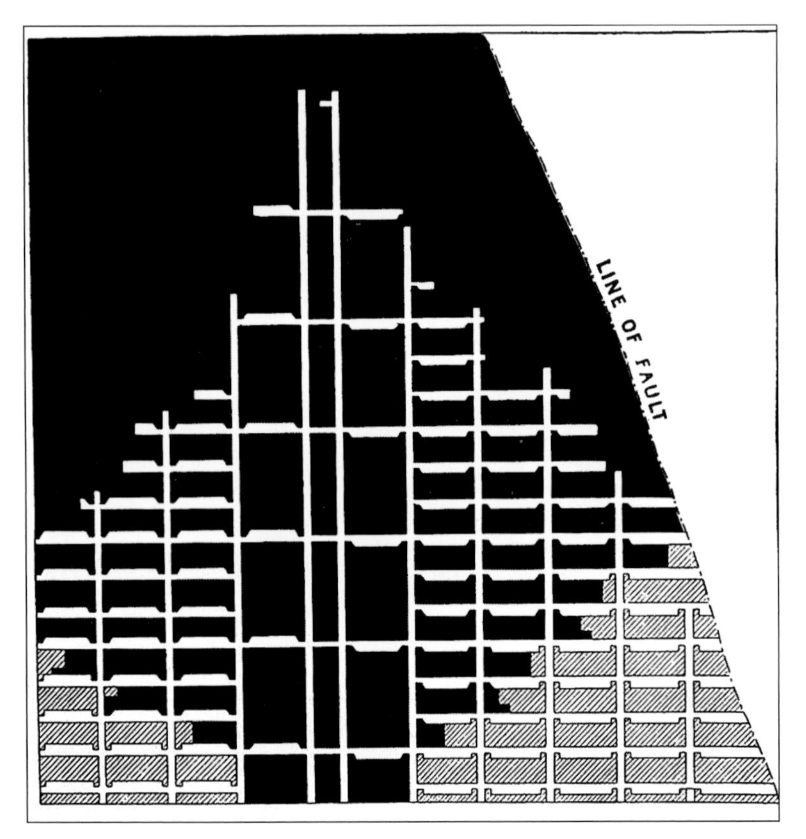

⬆ **A bord-and-pillar district. The size and shape of the pillars would depend upon the depth of the pit and geological conditions.** *(Author/Boulton)*

⬇ **Pillar working, with the seam divided into 'panels', sections separated by barriers of coal to minimise the risks of fire and explosion.** *(Author/Mason)*

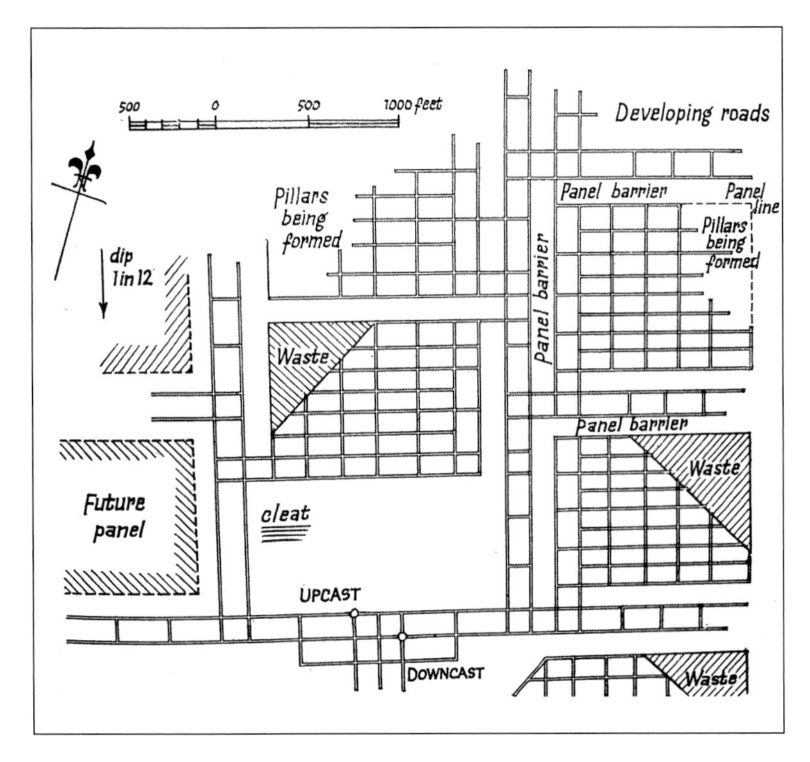

hose), all of which might be used to line mine roadways and other parts of the workings.

As already noted, mine shafts might have multiple purposes, and doing so required them to be divided into compartments called 'shaft sets'. The vertical divisions were called 'shaft guides', which were designed to maintain cages, skips and counterweights in their correct position as they moved up and down the shaft. There are two main types of shaft guide: 1) fixed guides, i.e. rigid structures made of wood (more historically), rail and steel; 2) rope guides, used with multi-rope haulage systems. Horizontal compartmentalisation was performed by 'shaft buntons', which also acted as supports and fixing points for shaft guides, pipes and cables. Note that the use of rope guides obviated the need for horizontal buntons.

ROOM AND PILLAR MINING

We now move on to the subject of mine layout underground, another of those topics within mining that contains all manner of complexities and exceptions. Here, we shall compress the topic somewhat for clarity.

Once mining moved beyond the bell-pits, and deeper shafts had been sunk, different methods of exploiting the underground coal had to be pursued. The first method evolved from as early as the 13th century in Europe, and was to produce the primary mine configuration in the UK until the expansion of longwall mining in the 20th century (see opposite). To exploit the coal from the bottom of the shaft, the early miners cut small, self-supporting tunnels called 'headings' horizontally into the coal seam. Multiple headings were cut out from the shaft to avoid the problems of ventilation associated with just one tunnel, and then headings were also connected roughly at right angles via other, wider tunnels, called 'rooms'. Repeating this process continually, especially when it became a formal and regular method, produced a working pattern that consisted of a grid composed of the tunnels and rooms divided by the thick pillars of coal left in place, these supporting the mine's roof. Thus the method came to be known as 'room-

and-pillar', although it has acquired several other titles according to regional preferences, including pillar-and-stall, stoop-and-room and bord-and-pillar.

Room-and-pillar mining, although of lesser importance in the 20th century, was still used in coalfields in the UK long after the nationalisation of the coal industry in 1947, especially concentrated in north-east England, and is still used to this day in the USA and elsewhere. It is mainly applicable to mines no deeper than 1,300ft (400m); below this depth, the coal pillars will likely start to fail under the geological weight. The method was also limited mainly to seams that were at least 4ft (1.2m) thick, to allow for the haulage systems that transported the coal back to the shaft. Room-and-pillar also had a limitation in that the pillars held large amounts of viable coal. To access this, therefore, the mine was worked in two directions. The first way – known as 'working in the whole' – was the initial process of cutting the headings and rooms out into the coal seam. During this phase, a total of 30–50 per cent of the coal in the seam could be won, depending on the dimensions of the pillars. Once the limits of the working were reached, the miners would then 'work in the broken', removing the pillars on the way back towards the shaft. This was performed by gradually slicing off portions of each pillar, while supporting the roof with timber props. Once the pillar had been totally removed, and the coal had been extracted, the wooden support could then be drawn out and the roof collapsed in a controlled fashion. Naturally, this process had to be performed with much informed caution, especially as the removal of the pillars would result in the weight shifting on to remaining pillars. Typically, the pillars needed to be extracted a row at a time.

LONGWALL MINING

Globally, and certainly in the UK coalfields, longwall mining came to be the most widely adopted and productive of the mining methods. Although its rise mainly occurred from the mid-19th century and rapidly during the 20th century, its principles were actually laid down in the 16th century; longwall working is first seen in Leicestershire in 1625.

The longwall principle, as the name indicates, develops a long exposed coalface – typically somewhere between 109 yards and 273 yards (100m and 250m) long – which can then be mined as a whole, advancing forwards through the seam. The system, depending on the mechanisation used, can be enormously productive. In the USA, for example, a single longwall face holds a record for 3.5 million US tons (3.17 million tonnes) of coal produced in a single year, about 20,000 US tons (18,134 tonnes) per day.

The layout of a longwall mine working is subject to local design requirements, but the basic principle is as follows. First, a pair of 'gates' or branch roads are cut out from the mine's main haulage route, these running in parallel about 39ft (12m) apart. Next, the two gates are connected at their ends by a further

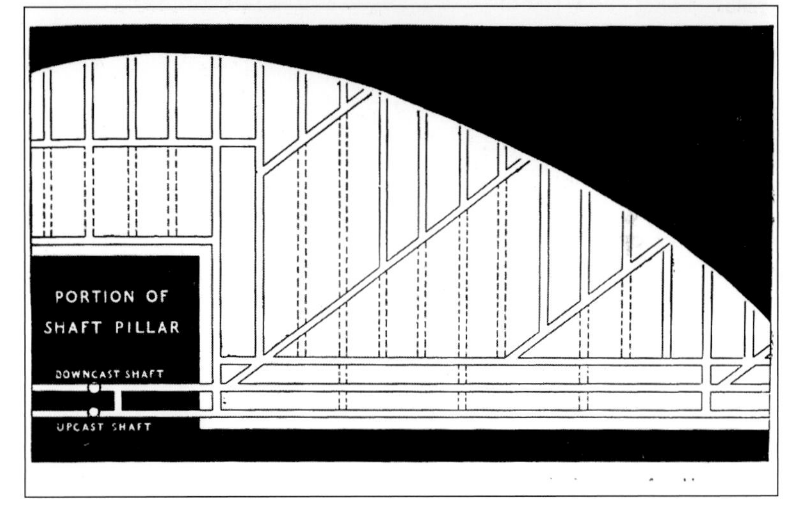

↑↑ Tower Colliery near the village of Hirwaun, in Glamorgan, South Wales, was the oldest continuously working deep-coal mine in the UK. *(Richard Whitcombe/ Shutterstock)*

↑ The working of a seam using the longwall method. Note how the downcast and upcast shaft provides circulatory ventilation.

LONGWALL MINING – ADVANTAGES AND DISADVANTAGES

Advantages

- Simple layout of the workings.
- Maximum extraction yield from the coal seam.
- Extracted non-coal minerals provide useful material for goaf packing.
- Convenient routes of ventilation.
- Roof weight pressing on the coalface helps loosen the coal.
- Ideally suited to mechanisation.
- Good for working beneath another worked-out seam.

Disadvantages

- Very high maintenance costs, particularly of the roadways.
- Risk of large-scale collapses at the coalface.
- The goaf can act as a reservoir for the build-up of methane gas.

Note that as the 20th century progressed, many of the disadvantages of longwall mining were mitigated. For example, roadway maintenance costs were reduced by retreat-mining roof-bolted roadways to leave the fractured strata behind (retreat mining is discussed below). The technique of advanced methane boring – whereby the methane was drained ahead of the face – was also implemented; at Tower Colliery, a fully functional methane boring power generation unit was installed on the surface to generate energy using this resource.

A 19th-century visualisation of longwall mining. The top part of the picture shows a perspective view of the men working the seam, while the bottom half shows the pit props and goaf in plan. 'A' and 'B' are ventilation shafts, and the arrows indicate air flow. *(Author/Tomalin)*

tunnel cut at a right angle to the gates. This tunnel exposes the face of the 'longwall' coal seam to be mined. Multiple gates connected in this way form the long coalface; once the coal is mined from the longwall, then the gates can be advanced again for the next cut of coal. The 'advancing' longwall method therefore left a large void behind it, known as the 'goaf' or 'gob', the area remaining after the coal was cut. This was supported with timber props in the immediate vicinity of the coalface, and further back by large permanent supports called 'packs' (see Chapter 5). The voids remaining in the goaf would be filled in by allowing the roof to collapse in naturally, although roads might be cut through the goaf to provide access to face workings.

Longwall mining can be approached in two ways. 'Advancing' longwall mining, already mentioned, begins in the vicinity of the shaft pillar (an enclosure of rock running around the shaft to provide additional structural support) and works outwards, moving the face forwards a step at a time. In 'retreating' longwall mining, the roadways are driven straight out to the very boundary of the intended workings, before being connected to form the face. The miners then work that face in the direction from which they have come, i.e. towards the shaft. This system does not need the goaf packing for roof support.

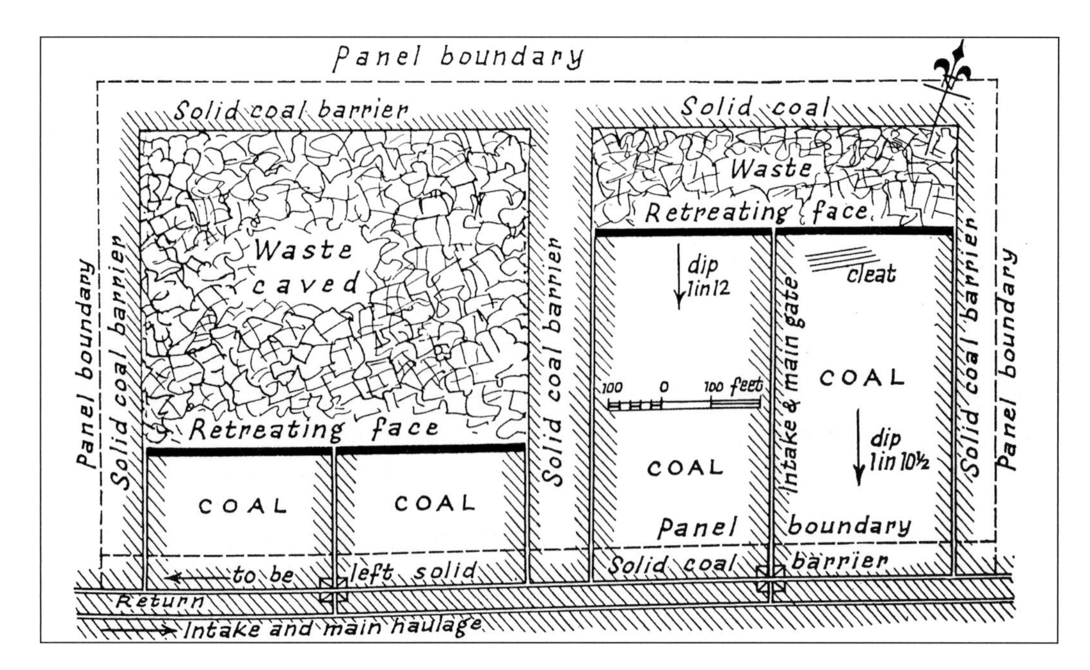

A longwall retreat mine, the face being worked back towards the haulage routes from the panel boundaries. *(Author/Mason)*

HORIZON MINING

Horizon mining has been a rarer method in the UK, but the system attracted the attention of the NCB in the aftermath of World War II and a handful of pits were developed using the system in the post-war era. Examples include Nantgarw Colliery in Rhondda Cynon Taf and Wyndham Western, Bridgend (both in Wales). The horizon mining method was mainly for use when working inclined or faulted seams. To access the seams, a series of main roadways were driven out horizontally or on a slight incline, through the measures or strata, the roadways being tunnelled out from the shaft at pre-arranged intervals of depth. Each roadway or level is known as a horizon, and they have been likened to arteries stretching out through the entire measure of coal. The horizons are linked by vertical tunnels. It should be said that horizon mining is actually not a specific method of 'winning coal' (the term used for its extraction), but is more about the arrangement of the mine; within the horizon mine, regular longwall or pillar-and-stall methods of working might be used.

THE MINE ABOVE GROUND

The hidden workings of the underground coal mine lay beneath the very visible industrial complex on the surface, which contained a host of buildings, often spread out over a site many acres in size and hugging the contours of the land. A coal mine was a vast work of human and mechanical engineering, and extracting the coal below ground was just the beginning of a long mining process that was continued above ground.

It should be said that colliery development and planning has almost as many variations as

Coal tubs wait at the pit head to take coal away from the pit top for processing. The profile of tubs varies according to local preferences and whether constructed of wood or steel. *(Author/Big Pit NCM)*

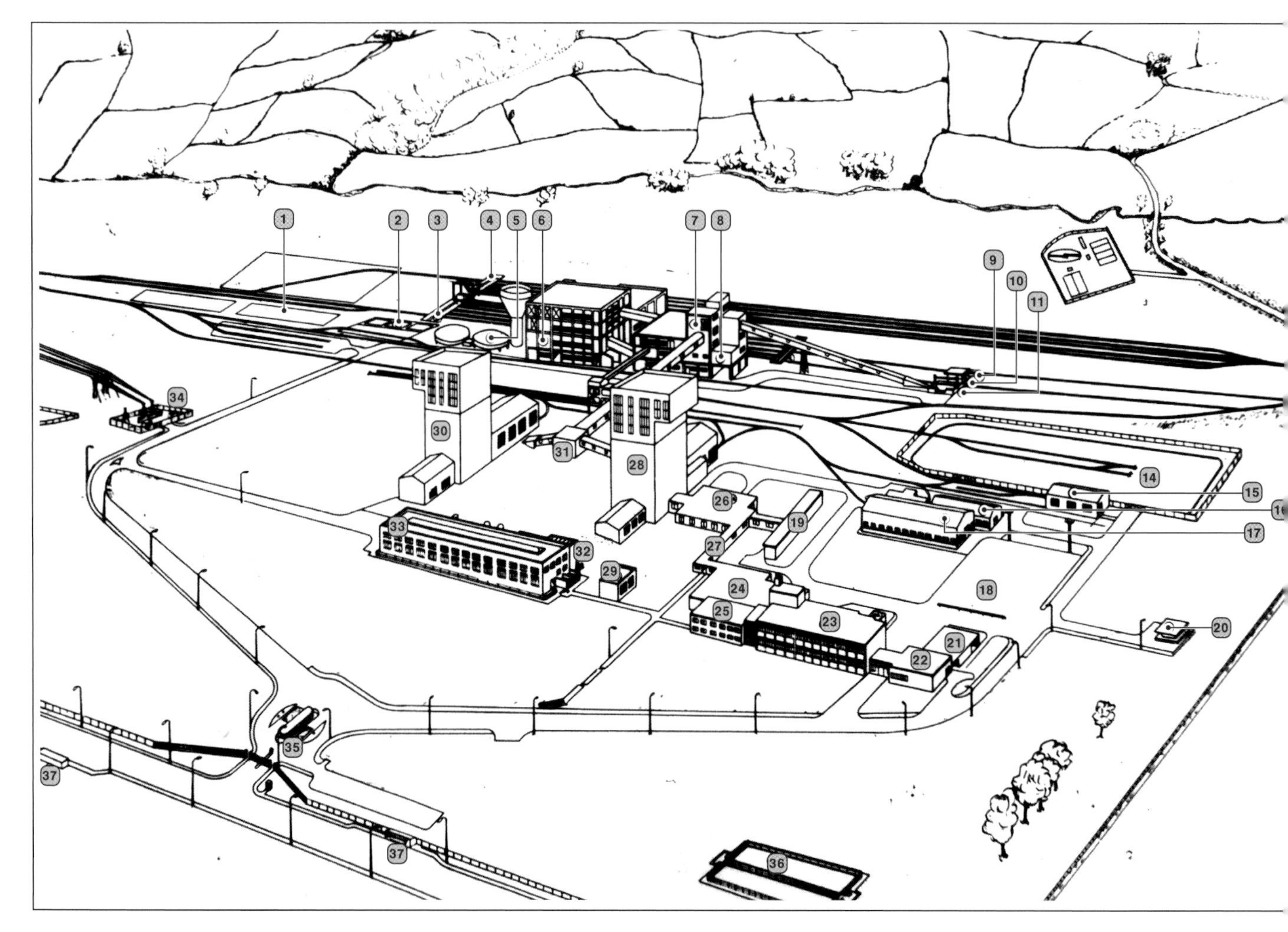

⬆ **A National Coal Board aerial plan of Abernant Colliery, cleariy showing the integration between the mine and outgoing transport links.** *(Author/NCB)*

⬇ **A mine fire station. By the 20th century it was law that pits with more than 100 employees had to conduct fire-fighting drills.** *(Author/Big Pit NCM)*

there are pits themselves. Throughout much of UK mining history, collieries were private enterprises, and as such were developed according to the priorities of the colliery management and the prevailing economic conditions. Much colliery development before the 1947 nationalisation was, therefore, evolutionary and frequently haphazard, the assortment of buildings and their functions changing with emergent technologies and financial fortunes. Nationalisation brought a more centralised and professional aspect to mine planning, and from the late 1940s to the 1980s the NCB rationally redeveloped many mines at great industrial expense. For example, the Maerdy Colliery in the Rhondda, South Wales, which had opened in 1881, received a £7 million investment between 1949 and 1959, transforming it into one of the most modern mines in the UK. Part of the reason for the redevelopment was to convert the mine to a horizon mining layout, which at this time

1 Coal stocking area
2 Settling ponds
3 Traverser (Fulls)
4 Traverser (Empties)
5 Thickeners
6 Coal preparation plant
7 Raw coal & stone prep plant
8 Stowage stone bunker
9 Weighbridge on tippler
10 Weighbridge
11 Foreign coal tippler and bunker
12 Railway bridge
13 Level crossing
14 Stockyard
15 Loco shed
16 General stores
17 Maintenance shops
18 Bus park
19 Powder issuing stores
20 Magazine
21 Cycle store
22 Canteen
23 Baths & lockers
24 Lamp room
25 Pit offices
26 Pit maintenance cabins
27 Covered walkways
28 Upcast shaft winder tower & car hall
29 Standby diesel alternator house
30 Downcast shaft winder tower & car hall
31 Transfer house
32 Ventilation fan drift évasée
33 Compressor house
34 Electricity sub-station
35 Road weigh house & fuel pumps
36 Water storage ponds
37 Bus shelters

⬆ A colliery blacksmith's workshop. Typical roles of the blacksmith included making, sharpening and mending mining tools. *(Author/Big Pit NCM)*

⬇ A sawmill was a standard building in coal mines, used for cutting timber to form props, facings and all manner of supports. *(Author/Big Pit NCM)*

was common in mainland Europe and was attracting the attention of the NCB. Given the number of pits in the UK, there was still much contrast to be found in the quality and logic of design and layout, but there were a set of common functions that needed to be housed at any coal mine.

As already discussed, the most prominent elements of the mine were the headframes, which have come to be an iconic representation of the coal mining industry in the round. The headframes would usually sit amid a complex of covered buildings that were associated with the immediate processing of the coal that had been drawn from the ground below. In these buildings, the coal might be screened for size, washed (washery buildings tended to appear at the end of the 19th century) and transferred into wagons, to be run down to nearby railway sidings, road networks or the quayside of a canal, for transportation onwards to the consumer.

Also in the vicinity of the pithead were buildings that contained the heavy machinery responsible for providing the core functions of a mine operation: winding, pumping and ventilation. Prior to mechanisation, the buildings or areas were largely spaces in which horse gins operated, or were constructed next to a flowing river to drive waterwheel-powered pumps. With mechanisation, however, the buildings were used to locate steam-driven, compressed-air then electrically powered machinery. Many collieries had specific houses for pumping

⬆ A colliery switchboard unit was essential to join together all the mine operations through coordinated communications. *(Author/Big Pit NCM)*

➡ The lamp room was the place in which miners' lamps were stored and recharged when not in use. *(Author/Big Pit NCM)*

Coal mines typically had integrated rail or canal transport links to take the mined coal direct to market. *(Author/Big Pit NCM)*

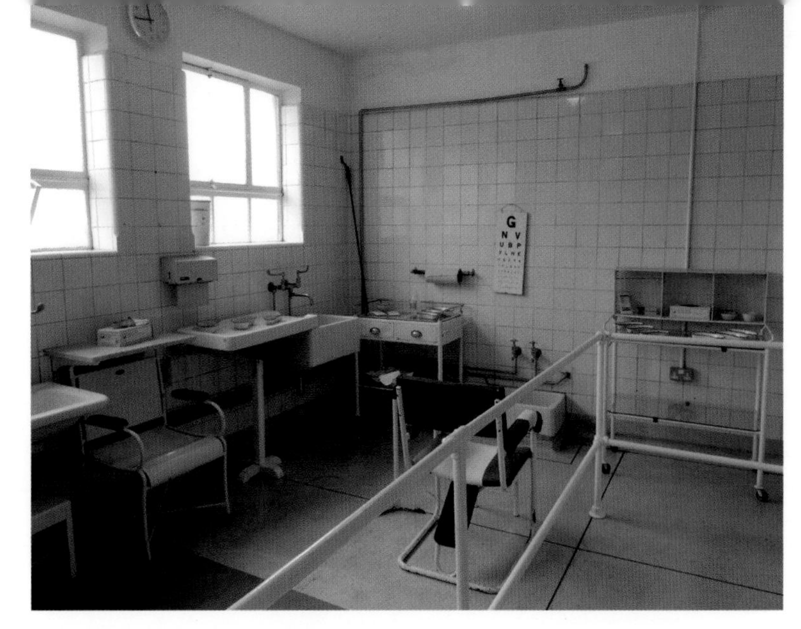

engines and winding engines, although in time it became common for both types of machine to be located in a single building. Prior to electrification, during the steam age, collieries might also have a specific boiler house, used to generate the steam that was then channelled off to all the machinery that ran on its power. The mine might in addition have its own reservoir, for holding the large amounts of water required to generate the steam. When electrification arrived in the late 19th and early 20th centuries, power buildings might house generators, transformers and compressors.

The requirements of below-ground ventilation were also apparent from the buildings around the pithead. Before the introduction of steam-powered then electrical ventilation fans, brick-built ventilation furnaces and chimneys were in evidence at many deep mines, replaced by fan houses for the new machinery from the first half of the 19th century.

The many mechanical and engineering demands of the colliery were served by numerous other buildings dotted around the pit site. These included blacksmith's shops, usually accompanied by stabling for the horses used in underground haulage. (There were still 21,000 pit horses in service even in the year of nationalisation.) Other workshops included foundries, engineering centres, fitting shops, sawmills, tool stores and explosives stores, although the latter were usually kept in barrel-vaulted brick storehouses in isolated parts of the colliery site, to reduce the damage inflicted by an accidental explosion. In addition to the workshop buildings, the pit was also home to a collection of office buildings, occupied by mine managers and those responsible for the colliery's administration.

A key building type to appear in the first half of the 20th century was the pithead baths, places in which the miners could shower, freshen up and change after a shift. The first of these was dated to around 1913, but it would take time and legislation before these essential facilities reached most miners. The pithead baths became far more common in the 1930s and 1940s, and not only housed the shower and washing facilities, but also came to include several other facilities, including boot-cleaning stations, medical centres (often with an accompanying morgue) and canteens. The pithead baths might also have the lamp room (the room in which the miners' lamps were kept and the batteries were charged) within its boundaries, although sometimes this was a separate building.

In many ways, the boundaries of a coal mine did not end at a perimeter fence or an outlying building. Many collieries nestled right in the heart of the civilian population they employed, the miners' homes and streets typically within walking distance of the mine in which they laboured. Looked at from a distance, a coal mine was truly more than the sum of its parts.

A colliery infirmary. Many miners would hold first-aid certificates, often gained through training via the St John Ambulance Association. *(Author/Big Pit NCM)*

In modern 20th-century mines, each miner had a 'clean' locker and a 'dirty' locker, storing his ordinary clothes in the 'clean locker' and leaving his work clothes in the dirty locker, where they would be dried by hot air blown through the lockers. *(Author/Big Pit NCM)*

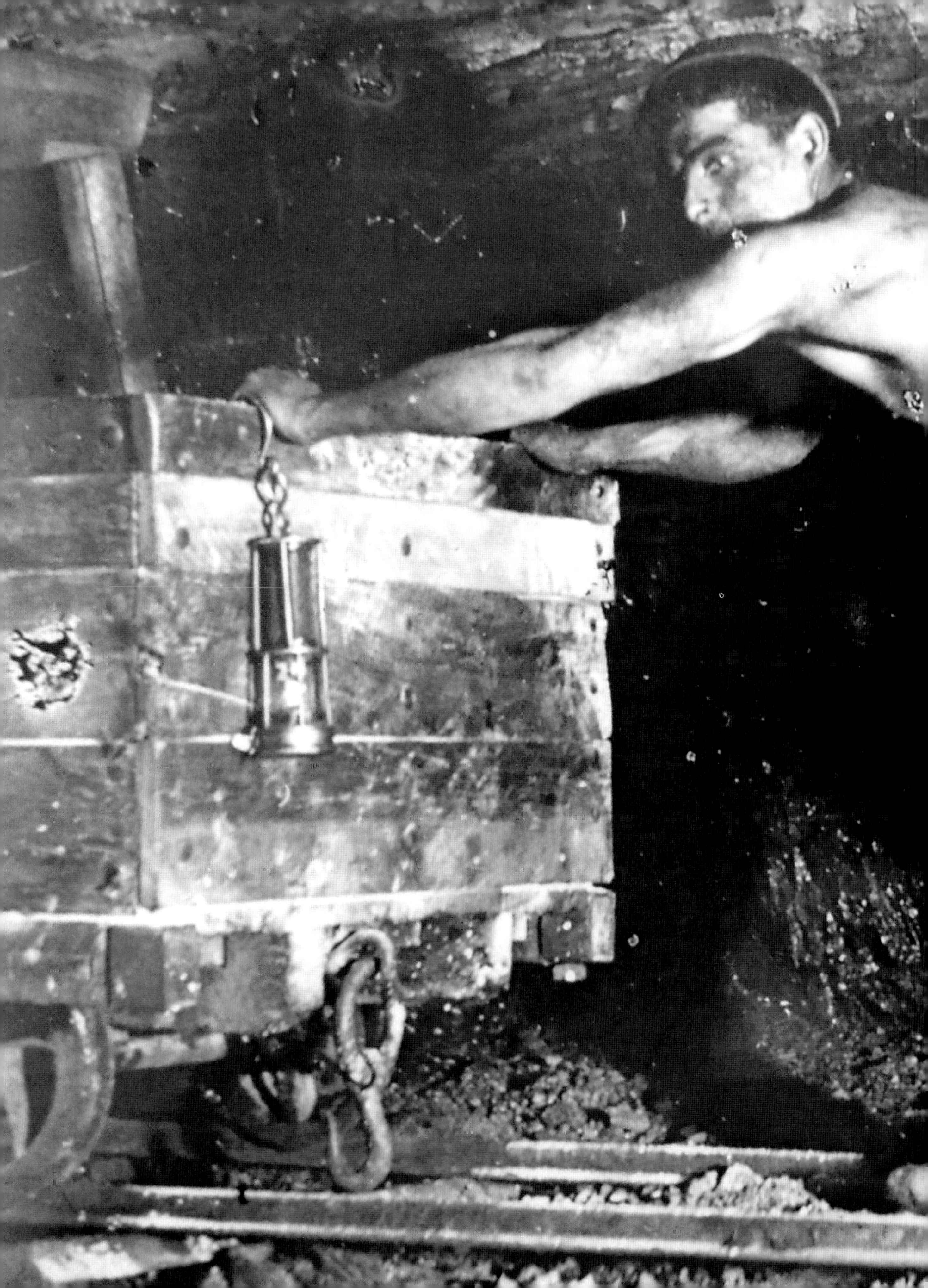

MANUAL COAL EXTRACTION AND HAULAGE

Throughout history, all deep mining has sought optimal interconnected efficiency in four key stages of the production process: 1) extracting the maximum volume of coal from the face in any given time (the time typically defined as a shift); 2) collecting the cut coal and moving it from the coalface to the mine shaft; 3) hauling the coal up the mine shaft to the surface; 4) unloading the coal at the surface and processing it for use or sale. Ideally, these four stages had to merge in a flowing sequence of movement; the faster the coal could be won, moved and processed, the greater the pit's potential output and therefore the better its economy.

⇦ Physical haulage of coal tubs was a feature of life in many small coal mines well into the 20th century. Over time, such effort could result in debilitating back, shoulder and hip injuries. *(Author/Big Pit NCM)*

Miners working a coalface with picks. Although this working area looks cramped, many faces would offer even less head room for the miner. *(Author/Big Pit NCM)*

MANUAL CUTTING

In the days before mechanisation, which really arrived in the 19th century, coal mining was overwhelmingly a job of brute human muscle power, aided by some basic tools and the occasional application of the laws of physics. Actually, the application of manual labour continued well beyond the advent of mechanisation, depending on the pit's modernity and the reality of the task at hand, so the descriptions of many manual techniques below, especially in terms of coal-cutting, should be regarded as tools in the box throughout mining history, not just in more distant times. For, even in the mid-1950s, some 10 per cent of British coal was still 'hand got', generally in cases where the use of machinery was impractical.

An important preliminary point to make in terms of winning the coal is that whatever cutting method was used, it could not be so aggressive that it pulverised the mineral into small pieces. Until the birth of industries such as electricity generation, which could use very small pieces of coal, almost down to dust, saleable coal had to be produced in large- or medium-sized pieces, to make it convenient to store, move and use by domestic, military and industrial customers. Coal is typically a frangible material, meaning that it readily shears, flakes and crumbles under pressure – another factor that the designers of the mining process had to take into consideration if it were to produce the right-sized coal for the market.

The manual process of coal mining was conducted with picks, hammers and shovels, and in its most basic outline followed a two-stage sequence:

1 Miners would first undermine a section of seam by digging a slot (a 'hole-out') at the bottom of the seam, just high enough so that he could wriggle into the slot lying on his side, while still wielding a pick. The hole-out would typically be cut to a depth

Until the mechanisation of the late 19th century arrived, coal-getting was a brute manual process. In this image, two young boys work together to break up rock with iron bars. *(Author/Big Pit NCM)*

of 6ft (1.8m), and to prevent collapse, short wooden temporary props called 'sprags' were driven up against the overhang.

2 Once the stage 1 process had been completed, the miners removed themselves from the hole-out, the sprags were knocked away, and picks and iron bars, assisted by the roof weight, were used to bring the section of seam down in a large mass for collection.

The undercut hole-out was not the only option for a cut into the seam. Cuts could be made in several other locations on the coal seam, including in the middle and at the top, and in vertical sections. The ability to make these cuts was largely a product of mechanisation, however, so will be studied in more depth in the following chapters.

Although this activity might sound relatively crude, it actually required a solid bedrock of experience and 'pit sense' – the ability to 'read' both the coal seam and the working environment – for the sake of safety and productivity. Before attempting to bring down a section of coal, the knowledgeable miner would often tap the surface of the coal, the audible feedback providing clues about its structure. He would make a close visual

A miner would use his pick to 'undercut' the seam, utilising the weight of coal above to bring large sections of coal down in volume. *(Author/Big Pit NCM)*

A miner undercuts a low seam in this photograph, likely taken at the end of the 19th century. The job would have been made more dangerous by the low illumination, here a naked candle. *(Author/Big Pit NCM)*

inspection, examining the coal's lamination and bedding (terms referring to the strata of the rock) and looking for its 'grain'; like wood, coal is more easily split along the grain than across it. He would also study the coal for 'cleats' – naturally occurring fault lines and fractures that can be exploited through pick blows to bring down the coal in a large mass. Armed with this information, a miner who was an expert in hand-got techniques could bring down hundreds of kilograms of coal with judicious pick strikes at key points.

One of the best descriptions of the intricacies of manual mining comes from the writer John Farey, who in 1811, as a member of the Board of Agriculture, published his monumental and catchily titled work *General View of the Agriculture and Minerals of Derbyshire; With Observations on the Means of their Improvement*. In a particularly fascinating section explaining mining practices in the Derbyshire coalfields, he outlines, with an unswerving avoidance of full stops, the coordinated efforts by which coal was cut by hand, and the specific roles of the miners therein:

> *'The working now commences, by a set of Colliers called Holers, who begin in the night, and hole or undermine all the bank or face of the Coal, by a channel or nick from 20 to 30 inches back, and 4 to 6 inches high in front, pecking out the holeing-stuff with a light and sharp tool called a pick, hack, or maundrel: and placing short struts of wood in such places where the Coal seems likely to fall, in consequence of being so undermined. On the facility of this holeing, much of the profit of the seam depends, as well as on the roof: in favourable case, a thin stratum of sloam, spravim or soft earth, or of soft and bad Coal, lies under the valuable seam, in which the holeing can be made, and a hard rock or bind covers the Coal, so that the whole of it can be got: in less favourable cases, the holeing is obliged to be made in the middle of the seam of Coal, where there happens to be a soft bat of earth or bad Coal, when the labour of wedging up the lower part and the breakage of the whole, is much increased; and in the worst cases, the holeing it forced to be made in good Coal, chipping and wasting it, and at the same time, some of the best Coal is obliged to be left for a roof to punch to, an operation that will shortly be mentioned.*
>
> *When the Holers have finished their operations, through the whole length of the Bank, or Banks, and cut a vertical nick at one or each end of the Bank, called the cutting end, and have retired, a new set of Men called Hammer-men, or Drivers, enter the works, and fall the coal, by means of long and sharp iron wedges, set into the face of the Coal at top or near it, according to circumstances, which they drive by large Hammer, till the Coal is forced down, and falls in very large blocks, often many yards in length: this being a very dangerous part of the operation in the first bank, and before there is room, as afterwards, to step back between the puncheons, when the Coals fall: a man called the Rambler next follows, and with a hammer-pick breaks the blocks of coal into sizeable pieces [for transportation away]...'*
>
> **(Farey 1811: 344–45)**

Each of the men to whom Farey refers would have been expert in their specific roles, understanding the coal and the seam and the

Despite technological innovations, coal mining often still involves heavy manual labour. Here we see miners at work in Donetsk, Ukraine, in 2013. *(DmyTo/ Shutterstock)*

best application of their tools. The 'Punchers' were men who inserted wooden roof braces called 'puncheons' against the roof of the freshly cut seam, the posts tightened against the surface by hammering in small, flat pieces of wood called 'templets'. Note also the considerations about the composition of the coal and surrounding strata. Typically, a high percentage of the material mined would consist of non-coal minerals, which would have to be screened out in surface processes, discussed in Chapter 4.

THRUSTERS AND TRAPPER

A 19th-century artwork showing 'thrusters' – those pushing the tubs – and a 'trapper', a boy responsible for opening the airway doors. *(Author)*

SHOTFIRING

A dynamic new method of coal extraction came in the 19th century, with the use of explosives to bring down the undermined seam. In outline, the process involved drilling multiple holes at strategically chosen intervals along the coal bank, and into these were inserted tubular explosive charges. When subsequently detonated, the explosives instantly fractured the coal, bringing it crashing to the floor in a heap, with a consequent reduction in manual labour and an equally consequent increase in productivity.

Given the atmospheric dangers of explosions within a coal mine, there were evident dangers in using explosives underground. Firedamp – the flammable gas produced by cut coal – when at the right concentrations for firing, can be ignited at temperatures of about 666°C (1,232°F); the explosives typically used in mines ranged from standard gunpowder with an ignition temperature of 315°C (600°F) through to 2,204°C (4,000°F) for high explosives. Explosives brought with them further dangers. There was the risk of accidental explosions, usually through human error and careless handling rather than the fault of the explosives themselves. In addition, explosives generate poisonous gases of varying concentrations and threat levels, harmful to human health if breathed in significant volumes.

Although I have placed our consideration

A mine official inserts an air sample into his safety lamp to check for the presence of 'firedamp' (methane gas). *(Author/ Big Pit NCM)*

MINING SHIFTS

Coal mining is traditionally conducted in 'shifts', discrete periods of labour time during which specific tasks are performed and which ensure the mine is running at full productivity around the clock. The *Deputy's Manual* of 1956, written by E. Mason as part of the *Coal Mining Series*, describes how the 24-hour day was divided into three shifts, related to 'preparation, filling and packing' (Mason 1956: 249). In this system (more specifically relating to a machine-cut face), one shift would be spent cutting, boring, blasting and preparing the coal; the next shift would be the 'filling shift', transporting the cut coal away from the face; finally, the next shift 'would be engaged in moving the face transport system, extending the gate conveying or loading arrangements, ripping, packing and drawing forward the breaking-off line in the waste' (Ibid). In contrast to this cyclical process, 'continuous mining' was also in use by the time Mason wrote his works. Made possible by new developments in mechanisation in the 1940s (see the following chapters), continuous mining involved all three processes taking place simultaneously, dramatically raising the output of coal per man hour.

⬅ **A shotfirer implants explosive charges into boreholes drilled into the coal seam, prior to blasting.** *(Author/Big Pit NCM)*

⬇ **A shotfirer prepares to detonate explosives by sending an electrical current to the detonator from his single-shot exploder device.** *(Author/Big Pit NCM)*

of explosive mining in the chapter on 'manual' coal-getting, explosives remain a fundamental tool of mining to this day, hence their application is also relevant to the age of mechanisation. It is highly specialised work, requiring much knowledge and long training to perform effectively and safely. The key considerations involved in the shotfiring process have changed little over time, and revolve around:

- the type of explosive required
- safe handling of explosives
- the placement and packing of the explosives
- the detonating mechanisms

TYPES OF EXPLOSIVE

By the end of the 19th century in the UK, explosives were, by various Home Office Orders and Acts of Parliament, separated into 'permitted' or 'non-permitted' types for use in mining. The permitted types were drawn up on a fairly extensive list; a Home Office Order dated to 17 December 1906 listed 58 permitted explosives. From the 1940s, those on the list came to be distinguished clearly by a large 'P', enclosed in a royal crown, printed on the explosive's wrapper, this meaning that the explosive was authorised by the Ministry of Fuel and Power. Many of the Home Office-permitted explosives were also 'sheathed', meaning that they were encased in an inert material, such as sodium bicarbonate, which

The bulk transportation of the explosives above ground was accomplished via specially marked and coloured tubs. *(Author/Big Pit NCM)*

served to reduce the levels of flame and hot gases when the explosive was detonated.

The NCB's 1952 manual *Handbook on Shotfiring in Coal Mines* neatly divides the range of explosives on offer into two categories: gunpowder and high explosives. Crudely speaking, gunpowder is a 'low explosive' that, when ignited, goes through a process of *deflagration*, meaning that the flame front (the edge of the combustion) travels at subsonic speeds. High explosives, by contrast, *detonate*, the process of their decomposition producing a supersonic shock wave that is usually measured in several thousand metres per second. The *Shotfiring* manual further subdivides high explosives into 'gelatinous types' (containing a high proportion of nitroglycerin) and 'powder types' that only contained a low proportion of nitroglycerin.

The subdivisions of explosives might sound arcane, but it had absolute relevance for the application of explosives to coal mining. Some of the points made in the manual, and in other sources of the early 20th century, regarding this matter include:

- Gunpowder's relatively low explosive force made it useful for fracturing rather than shattering soft coal, but it was less effective in hard coal or hard rock.
- Gelatines with a high nitroglycerin content were the most powerful explosive types, and were therefore best used on hard rock or hard coal.
- Gelatines with a low nitroglycerin content were better for moderately hard rock or hard

A typical explosives store, kept physically distant from all other workings to reduce the risk of casualties should there be an accidental explosion. *(Author/Big Pit NCM)*

coal, although they could also be used on soft coal.
- Powder explosives were hydroscopic (meaning that they absorb water), so in water-logged seams gelatine explosives were to be used.

These points, and myriad other rules, made explosive selection a critical decision. Get it wrong, and tonnes of valuable coal could be turned into a useless dust, or the blasting might fail to bring the coal down in the first place. The objective was to fracture and dislodge the coal, not blow it up.

SAFE HANDLING

Safe handling began with storage. As noted in the previous chapter, explosives and detonators were kept in a dedicated storehouse in a remote location at the pit. Acts of Parliament in the 19th and 20th centuries laid down strict regulations regarding the structure and organisation of these buildings. The layout of the buildings ensured that different types of explosives and the detonators were held in their own areas, separated by thick brick walls, steel doors and spaces between compartments to absorb blast. Anyone entering the building had to don woollen overshoes, and any metal objects that could generate sparks – including hobnailed boots – had to be left outside. The explosives stored inside the building would be regularly inspected for damage or deterioration.

Explosives and detonators were transferred into the mine in heavy, lidded containers, ideally self-locking. The materials used for these containers included tin plate, leather, wood, moulded rubber and galvanised iron. In many ways, the most dangerous components carried by the shotfirer were the detonators (the mechanisms inserted into the explosives to trigger the explosion of the main charge). These were small but volatile mechanisms, prone to detonation through being knocked or crushed. According to the *Practical Coal-Mining* manual of 1908, 'The chief risk lies in their getting dropped about the mine, and filled up with the coal' (Boulton 1908: 237). It was not uncommon in the heyday of coal for a

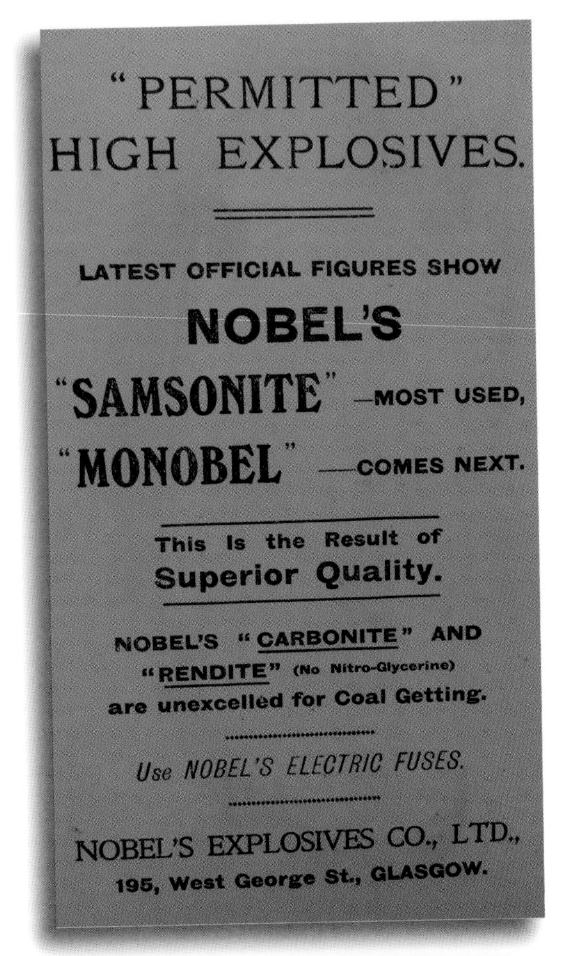

Hundreds of tons of explosives were used every year in the coal-mining industry. Here an explosives company promotes its 'permitted' products. *(Author/Big Pit NCM)*

Dummy explosive charges. Underground, explosives were typically carried in cylindrical metal canisters (as seen here) designed to hold no more than 5lb (2.27kg) of charges. *(Author/Big Pit NCM)*

domestic fireplace to be destroyed in a small explosion from a lost detonator making its way into the home.

PLACEMENT AND PACKING

While explosives might appear efficiently powerful, solid ground is actually quite capable of absorbing a great deal of an explosive shockwave with little visible effect. Furthermore, as noted earlier, the object of the shotfiring is to bring down the coal in large slabs and lumps, not to pulverise it. For these reasons, shot placement was key to achieving the best results from the firing.

The chief considerations in shot placement were (and remain to this day):

- the thickness of the seam
- variations in the hardness or softness of the seam
- the position of the cut in the seam
- the direction of the main cleat in the coal
- the type of roof and floor

The last point in this list was particularly important, because if the explosions damaged the roof unduly, it would become more unstable and difficult to prop, and if the floor were damaged it could make the collection of the cut coal problematic (it is always easier to scoop and load coal from a flat surface) and again interfere with propping. The cut in the seam was also integral to the blasting outcome, as it provided a crucial 'second face'; explosives placed straight into an unbroken or uncut face of coal will actually have little effect, whereas the undercut provides a space into which the coal can collapse.

The cylindrical explosive charges were set into specially bored holes, just greater than the diameter of the charges themselves. These were drilled by hand (a steel rod hand drill hammered manually into the coal), by a rotary hand-operated drill or by powered percussive or rotary drills (compressed air or electricity). The depth of the drillings varied from a few inches to several feet in depth, and they might also be angled upwards or downwards to increase the depth of coal affected by the blast.

While a single 'shot' (unit of explosive) might be used for a small-scale blasting, in most cases multiple shots were placed across the seam, in all manner of configurations. They might follow a straight line or a zigzag pattern, depending on the depth and nature of the seam. If the seam contained variable strata of material, more shots would be concentrated among the harder coal and stone, and fewer in the soft coal.

A multi-shot exploder, used for firing many charges at once or in a rippled sequence. *(Author/Big Pit NCM)*

FIRING

Shots could be fired either by a lit burning fuse or by electrical firing. The former sounds primitive, but can still be found in use; the fuse is made from a core of gunpowder or similar composition, with a flame-retarding covering that reduces the burn rate to 80–100 seconds per metre – plenty of time for people to get to positions of cover.

Electrical firing became the preferred method, however, during the 20th century. The detonators act as the interface between the electrical signal from the firing unit and the explosive. In mining, the detonators can be of either instantaneous or delay types. Instantaneous detonators trigger the explosion at the very moment the electrical signal to fire is given, whereas delay detonators impose a pre-set delay. The combination of the two is useful when firing multiple shots, to give a ripple effect that more finely controls the displacement of the coal.

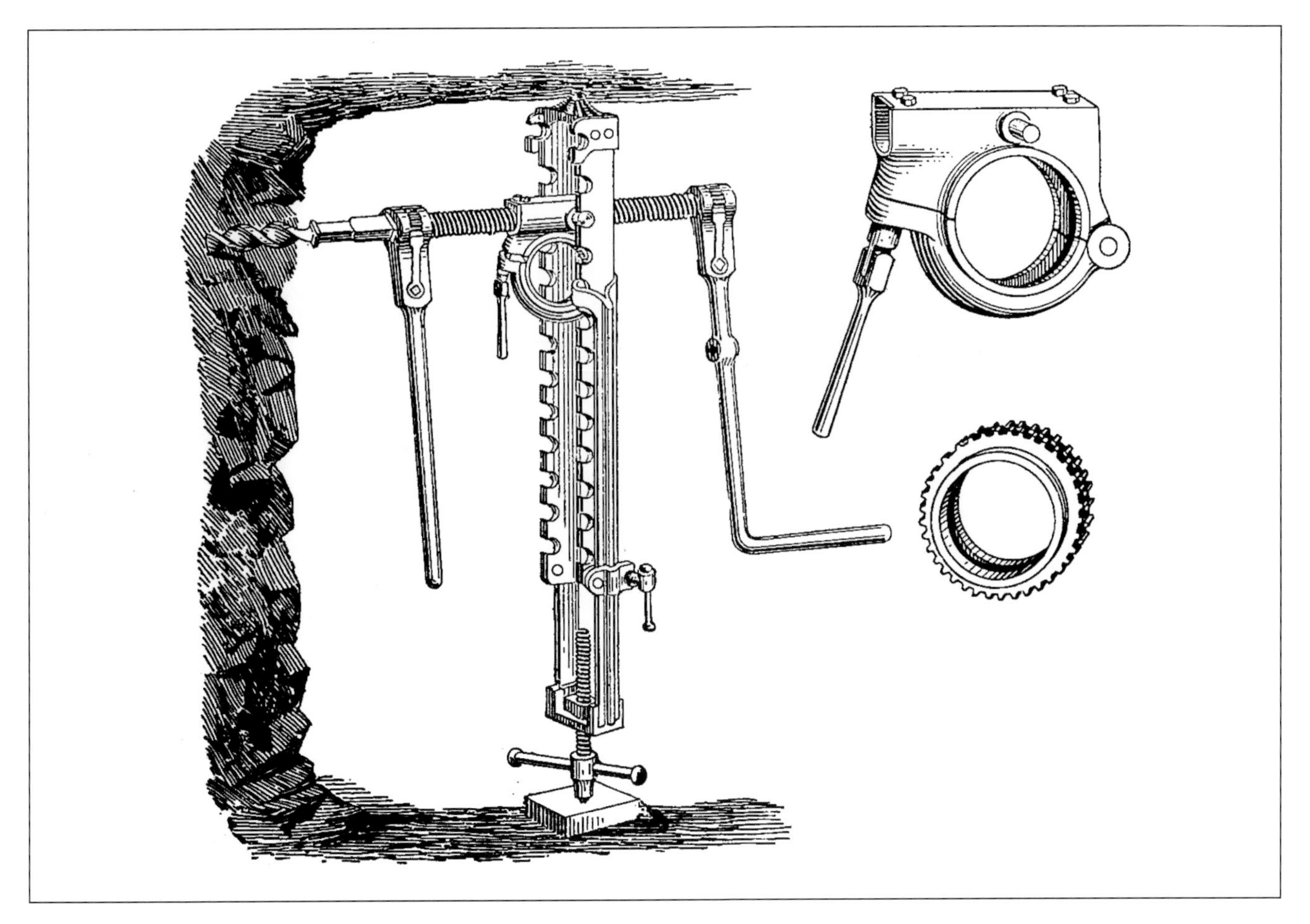

⬆ **The Elliot Machine was a drill mechanism used for hand-boring blasting holes into hard stone.** *(Author/Boulton)*

MANUAL LOADING AND HAULAGE

Cutting coal was just the beginning of the mining process. The loose coal now had to be loaded, moved from the face to the gateway, hauled up the gateway and the road back up to the shaft, then lifted to the surface. The challenge of this operation cannot be underestimated. Remember that both longwall and pillar-and-stall workings might have dozens of active faces, and tunnel networks could be measured in tens of miles, typically feeding to only one or two lifting shafts. If the coal could not be transported quickly and in coordinated fashion between face and surface, at best it would limit the productivity of the pit, and at worst it would bring the coalface operations to a complete stop for significant periods.

In this chapter, we will look at the human- and horse-powered methods of haulage, which persisted throughout the 19th century

⬇ **Miners load tubs via hand and shovel, early 20th century. The miners had to work carefully to avoid breaking up valuable large lumps.** *(Author/Big Pit NCM)*

A 19th-century illustration shows women employed in the back-breaking job of carrying corves of coal to the shaft bottom. *(Author/Big Pit NCM)*

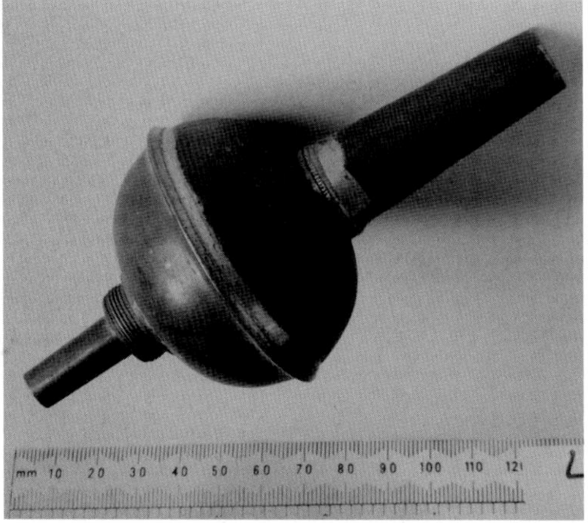

A 'peg and ball' oil lamp of a type common in Welsh mines during the 19th and early 20th centuries. The bulbous section was the oil reservoir. *(Author/Big Pit NCM)*

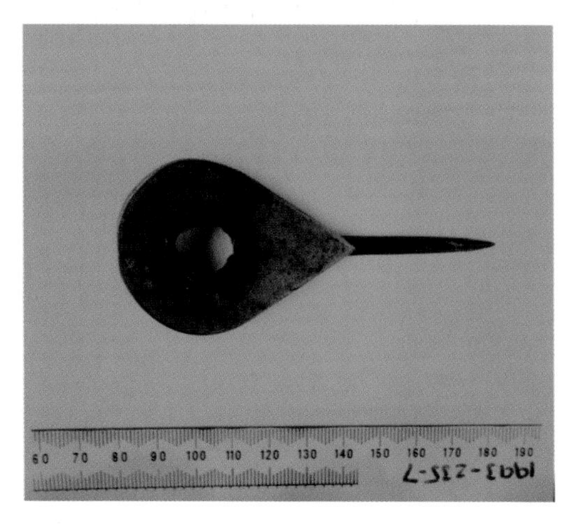

A miner's candle spike. The aperture held the body of the candle, and the spike was driven into wood or a crack in the stone to give the miner crude hands-free illumination. *(Author/Big Pit NCM)*

and even penetrated into the 20th century; steam-powered and more advanced processes of mechanisation are covered in the following chapter.

CORVES

The earliest coal haulage system, used primarily from the 16th to early 19th century, was the 'corf', a hazelwood wicker basket into which the coal was shovelled and then bodily dragged or carried to the shaft and the surface. The capacity of a corf is given, in a 19th-century dictionary of mining terms, as '10 to 30 pecks', a peck being an old imperial unit of dry measure (1 peck was equivalent to 2 imperial gallons/9 litres), so the weight of an individual corf must have been formidable. Those responsible for hauling the corves to the shaft and surface were paid less than miners (despite the importance of the job), and thus, given the times, women and children formed a significant part of the mine haulage workforce. A historic illustration shows a woman bent over at a right angle with the corf strapped to her back, but even this cramped position was only possible if the height of the roadway allowed it. Given that the tunnels around the mine might, depending on the pit, actually be little more than the height of the coal seam, corves would often have to be dragged behind the person, who literally crawled their way forwards through the tunnels on all fours.

The use of corves persisted into the 19th century; the dictionary of mining terms, published in 1888, gives a sense of the chronology when it states that: 'Since the introduction of tubs about 50 years ago the use of corves gradually ceased.' Unevenly distributed improvements of sorts – at least improvements in coal haulage efficiency, if not the conditions of human labour – were actually made earlier than the 1830s. 'Trams' were introduced – essentially wooden sledges on to which the corves could be placed and, once again, dragged to the shaft bottom. These saw a reduction in friction over hauling the corf itself. If the height of the roadway allowed, the trams were given wheels. We might also have seen wheelbarrows running in such mines, a single-plank track laid on the floor to assist the passage of the wheel over the rough ground.

Found in a Cumbrian mine, this corf illustrates one of the most basic methods of manual coal haulage. *(Geni/CC-BY-SA 4.0)*

EARLY WHEELED HAULAGE

A 16th-century German mining manual entitled *De re metallica* (On the Nature of Metals) shows some of the earliest examples of four-wheeled tram-like haulage. The wheels ran on two tracks made of planks, with a narrow groove in between the tracks. A guide pin projecting down beneath the front wheels of the tram locked into the groove, ensuring that the tub stayed on course. The tram and guide-pin system migrated into British coal mines, not least because German mine experts and miners were often employed in British pits.

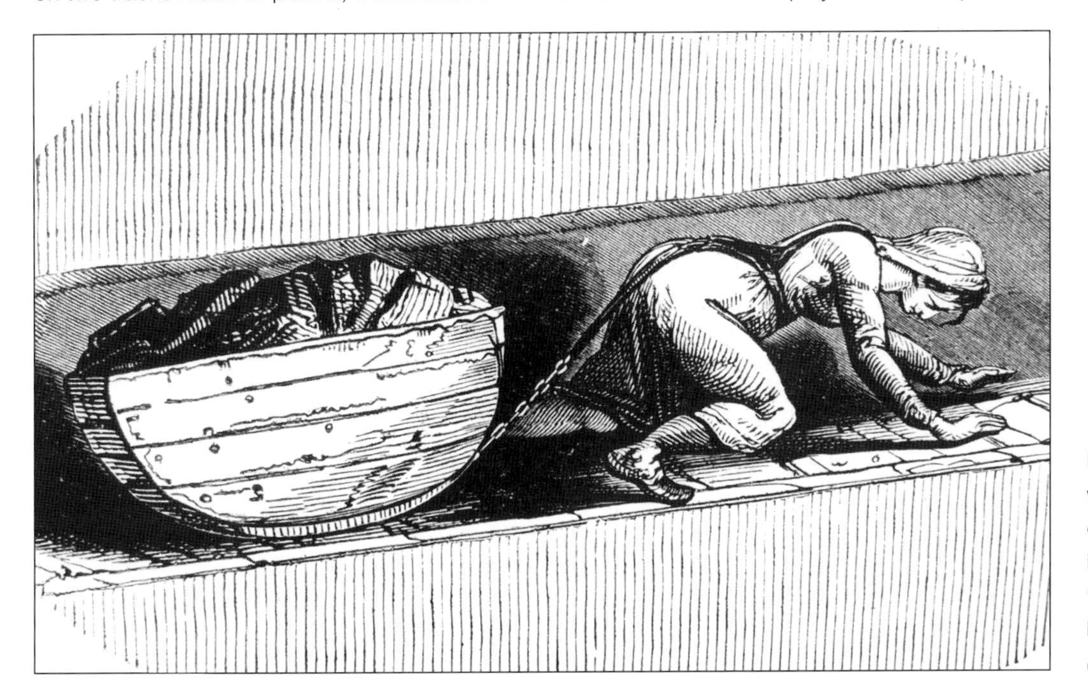

Manual coal haulage was made notionally easier through the introduction of the early 'tram' sledges. Note the basic wooden roadway. *(Author)*

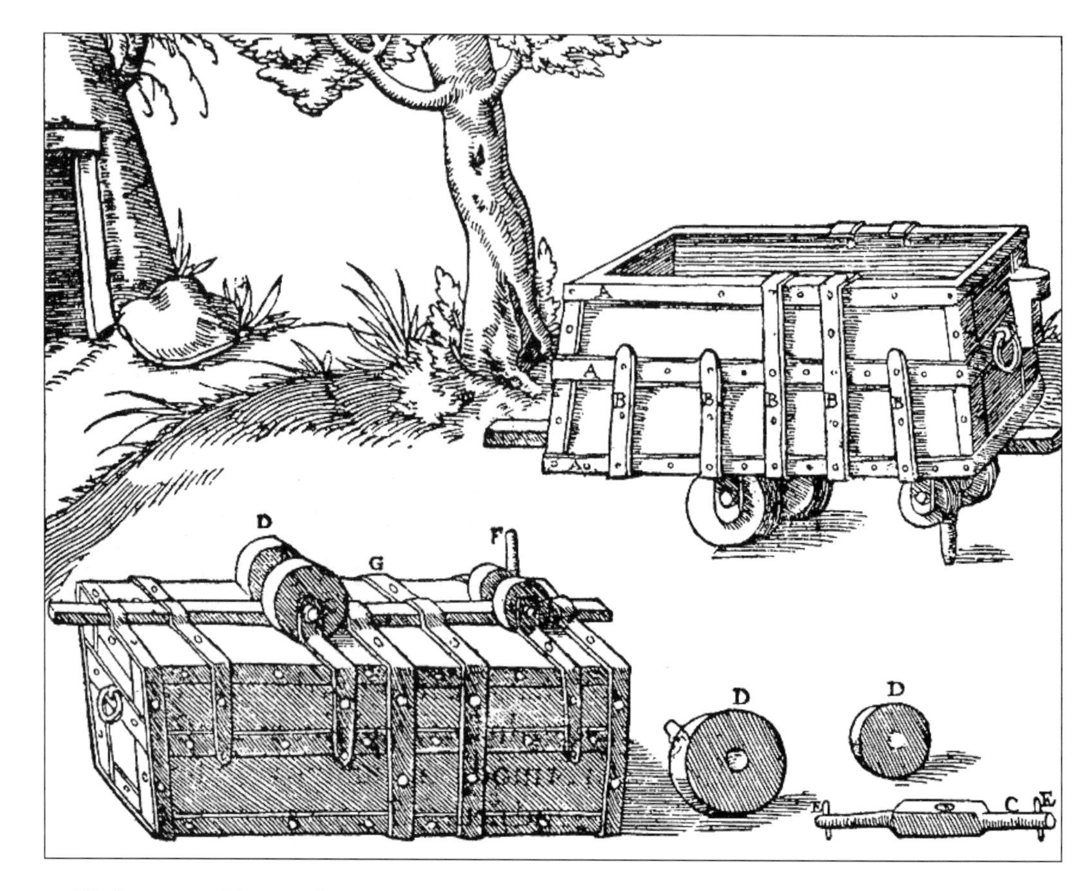

A 16th-century artwork of an early wheeled coal truck. A = rectangular iron supporting bands; B = iron straps; C = iron axle; D = wooden rollers; E = iron pin; F = downward projecting pin, which locked into a guide track between the wheels; G = same truck upside down. *(Georgius Agricola/PD)*

This child is using a waist belt to pull a coal tub through a confined shaft. Women, including pregnant women, also had to use this infernal device. *(Author)*

We first see evidence of more advanced four-wheeled haulage in the early 17th century, in Nottinghamshire, but it spread across the country throughout the century. Not that this improved the lot of the workers greatly. The wheeled corves, or solid wooden 'tubs', were still largely manhandled physically by humans, including children. The person pulling the tub from the front was known as the 'hurrier', and was chained to the wheeled contraption via a harness. Assisting her might be two 'thrusters' (often children), who pushed the tub from the rear. Working in 12-hour shifts, their lives were tragically hard. The use of children and women

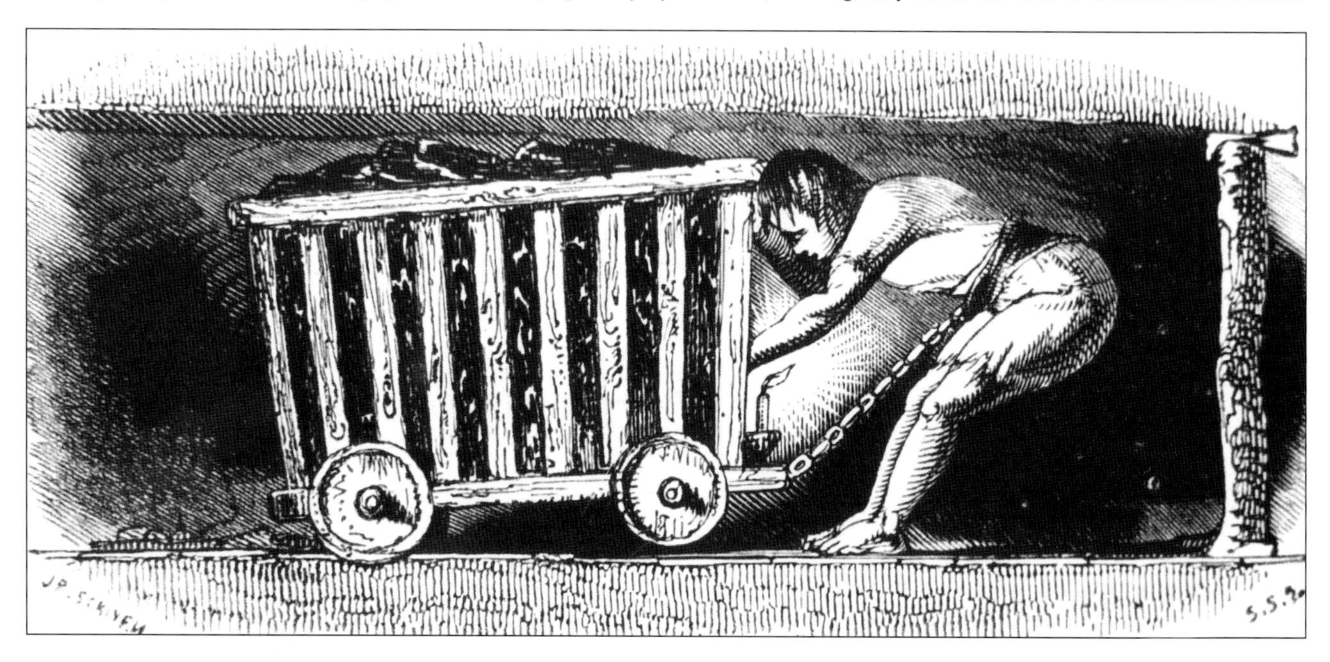

in coal haulage continued into the second half of the 19th century, when more humane legislation gradually restricted child and female employment down the mines.

An example of an early steel coal tub, with flanged wheels to keep the tub on the track. *(Author/Big Pit NCM)*

ROLLEYS, TUBS AND WAGGONS

During the 17th and 18th centuries, the systems of haulage were refined in some British mines. Corves were loaded on to a 'rolley', a wheeled horse-drawn carriage on which multiple corves could be placed and drawn. The efficiency leap was significant. In the rolley's original format, with wooden wheels that ran on basic rails, a horse could pull two or three rolleys, each holding two or three corves. Later improvements in the rails and wheels (particularly the substitution of wooden for iron wheels) meant that a powerful draught horse could draw as many as seven rolleys, similarly loaded, in a single journey.

In the 19th century, the wheeled tub, or waggon, became the preferred alternative method of haulage to the rolley. (There is some variation in the term used to refer to this piece of equipment; in South Wales, for instance, they were referred to as 'drams'.) The waggons were made of wood, iron or steel, the metal versions being sturdier and capable of holding heavier weights, although the wooden ones were cheaper and easier to replace and repair if damaged. The waggon system was much enhanced by developments in haulage rails. At first, in the 18th century, the tubs ran on planks overlaid with flat iron plate rails. In time, the side of each rail was bent up to form a flange that guided the iron wheels of the tub more reliably than did the old guide-pin system. Multiple tubs, each holding about a hundredweight of coal, could be linked together to form a 'train' – up to 20–25 if the roadway surface was a good one – and pulled by a sturdy horse. (Children were still often used to guide the horses.) When the tubs arrived at the shaft cage, they could simply be rolled into the appropriate cage compartment and hoisted to the surface.

A trapper opens the ventilation door to allow a man pushing a rolley to move through on his journey to pit bottom. *(Author/Big Pit NCM)*

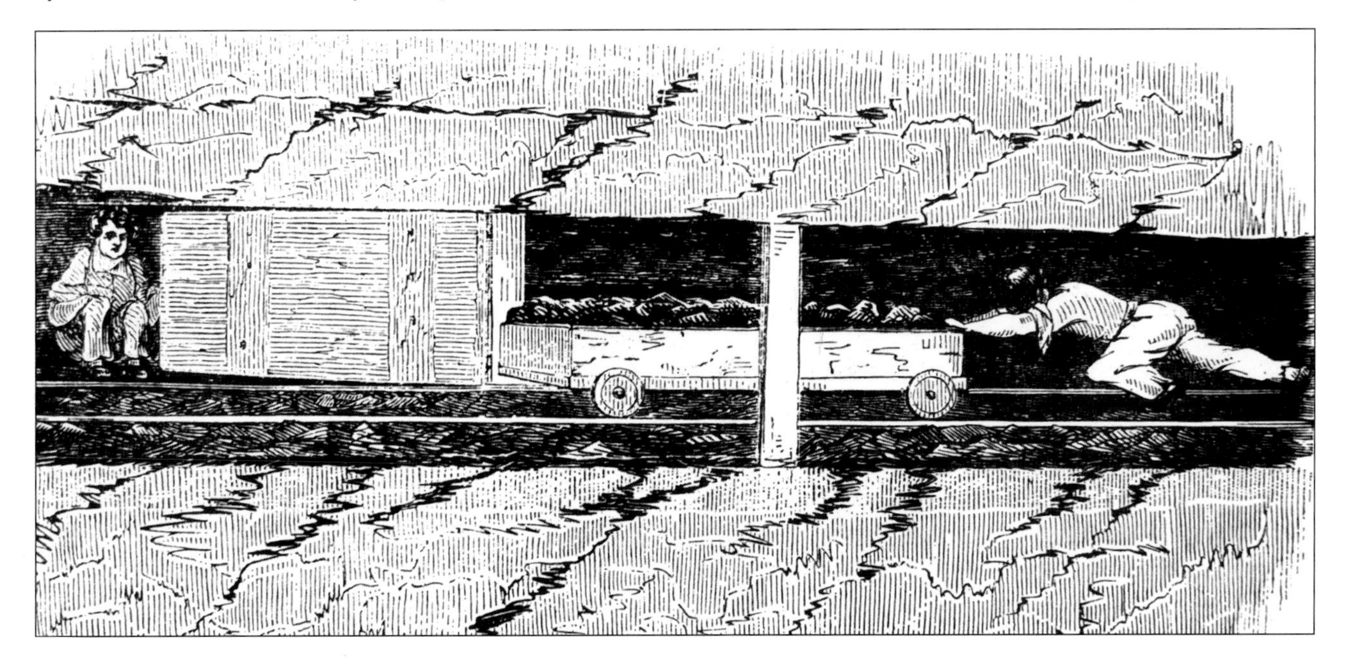

⬆ **The legal requirement for proper pit-head baths was a godsend for dirt-encrusted miners at the end of their shifts.** *(Author/Big Pit NCM)*

⬇ **The confines of the coalface often made the hand loading of tubs a physically awkward business.** *(Author/Big Pit NCM)*

ENDLESS-ROPE HAULAGE

Regardless of the quality or the resilience of the human or animal workforce, it was undoubtedly the case that muscle power was a limitation on coal production, and especially haulage. As coal-winning techniques improved in the 18th and 19th centuries, there arose the problem that coal was often extracted faster than it could be hauled to the shaft. This challenge would eventually be solved by powered mechanisation, but there were some earlier efforts to give the haulage process more physical assistance and efficiency. One of these used gravity and basic physics to its advantage, but was only suited to mines whose roads sloped upwards inbye (going away from the pit shaft towards the coalface) and downwards outbye (going towards the pit shaft from the coalface). With an 'endless rope' (meaning that the rope was formed into an unbroken loop) and pulley system installed at both ends of the roadway, a full load of waggons heading outbye could largely run with gravity assistance, while on the opposite side of the rope the weight of the loaded waggons pulled empty waggons back up to the coalface. The true revolution in rope haulage, though, would only come with mechanisation.

RAISING THE COAL

Going back to the days of the bell-pit and the earliest underground mines, the action of raising the coal up the shaft to the surface rested upon the shoulders of the miners, quite literally – the workers would climb ladders with the corves still strapped to their backs. This system had transparent human limitations and dangers. As noted in Chapter 1, therefore, the coal lift capacity was increased by a basic rope windlass system at the top of the shaft, which came to be improved and augmented by literal horse power. The 'cog and rung gin' ('gin' is short for 'engine') was actually a borrowing from agricultural practice, these machines used to perform tasks such as threshing or raising water from wells. It worked as follows. A lantern wheel in the vertical plane

A whim gin in operation, this one seen at the Beamish Museum in County Durham, England. *(Geni/CC BY-SA 4.0)*

HORSE WORKING LOAD

The 1888 publication *A Glossary of Terms used in the Coal Trade of Northumberland and Durham* by G.C. Greenwell, under the entry for 'rolley', included an interesting calculation about the weight of coal that could be pulled on a rolley by colliery horses:

'The following is given as the regular daily work of 12 hours of a good pit horse, upon a level rolley-way, in a good state –

EMPTY LOAD.

9 rolleys, weighing 7½ cwt. each, 27 journeys of 500 yards, equal to 204,120, led 500 yards.
18 empty tubs, weighing 2½ cwt. each, 27 journeys of 500 yards, equal to 136,080 yards

FULL LOAD.

9 rolleys, as above, 204,120 yards
18 full tubs, weighing 9½ cwt. each, distance as above, 517,104, led 500 yards

Total weight equal: 1,061,424, led 500 yards.

And taking the fraction at 1.130th part, or one half of that of common tub-way [. . .] the power of the horse is found to be for 8 hours equal to 25,515 lbs. raised one foot high per minute.

According to Desaguliers, a horse drawing a weight out of a well over a pulley can raise 200 lbs. for 8 hours together, at the rate of 2½ miles per hour, equal to 44,000 lbs. raised one foot high per minute. Mr. Smeaton states the efficiency of a horse at 22,000 lbs. raised one foot high per minute. The ordinary estimate of engineers is 33,000 lbs.'

An example of a surviving German whim gin. Note how the horse's winding track is under cover. *(Norbert Kaiser/CC BY-SA 3.0)*

was attached to the end of the windlass drum, and was driven by a cogwheel in the horizontal plane. The cogwheel was driven by a long wooden arm powered by one or more horses, which steadily plodded around at a speed of *c.* 2mph (3.2km/h). The horses thus applied their motive power to the windlass, and were therefore able to draw heavier weights of coal directly up the shaft.

The problem with the horse-powered windlass was that the presence of the mechanism and the animal at the top of the shaft limited the working room available for handling the coal as it arrived at the surface. For this reason, the cog and rung gin gave way to the 'whim gin'. Here, the horse-powered winding drum was set well to the side of the shaft head on a vertical spindle, and the ropes were run across to the shaft via a system of pulleys. The whim gin was the standard method of coal-raising in underground mines for more than a century. Whim gin mechanisms were located in special covered winding houses, to protect both the mechanism and the animals from the elements. In terms of lifting capacity, the author of an article in the *Compleat Collier* in 1708 calculated that a whim gin with a team of eight horses, working in shifts of two horses per day, would be able to lift 126 tons (128 tonnes) of coal per day from a shaft 360ft (106m) deep.

In this 19th-century illustration, a windlass is used to draw a coal tub up a steeply inclined roadway. *(Author/Big Pit NCM)*

PIT PONIES

The contribution of pit ponies to the history of coal mining cannot be underestimated. In 1951, more than a century after mechanisation had begun to transform the operation of coal mines, there were still 15,500 ponies working in coal mines in the UK. This was a major reduction from 1912, when 70,000 ponies were in mining service, and we can only guess at the numbers used in earlier centuries – estimates place the figures in excess of 200,000 in the late 19th century.

⬆ **A pit pony draws its load on the surface. Pit ponies were a feature of British coal mining well into the second half of the 20th century.** *(Author/Big Pit NCM)*

As a testimony to the efficacy of these hardy creatures, in 1992 there were still 24 pit ponies in service in Ellington Colliery, Northumberland, used to salvage equipment (RCHM 1994: 43).

Ponies and horses had applications to mining from the earliest days, not only pulling the wagons of extracted coal, but also operating the 'horse whims' that worked the mine's winding and pumping gear. The introduction of pit ponies underground, or at least the first recorded use, was around 1750 in the Durham coalfield, but the numbers of animals increased significantly during the 19th century, when legislated reductions in the use of child workers and women underground – especially the 1842 Mines Act – meant that pit ponies often provided the replacement labour.

Not all ponies were suitable breeds for mine work. They had to be strong, hardy, compact and of even temperament, not given easily to panic. The 1952 NCB manual *Pit Pony* explained that Shetland and Welsh ponies, which had a height of 9–12 hands (one hand = 4in/10in), were ideally suited for use in the thin coal seams of northern England, whereas the thicker seams of the Midlands meant that Dales ponies and Welsh cob breeds could be worked, these measuring up to 14 hands. The largest Welsh cobs or shire horses, up to and exceeding 15 hands, were used in the South Wales coalfields.

Up until the 20th century, the life experience of a pit pony was generally defined by the working practices, and the humanity, of the pit personnel who oversaw and operated them. It could be a cruel existence – endless hours of overwork pulling tonnes of coal, with injuries and ill health common, the horses often dying well before their time. Mining disasters such as explosions also killed ponies as they did men.

Legislation to safeguard the welfare of the pit ponies begins to emerge in the 1880s, but it was the Coal Mines Act of 1911, and the subsequent amendments in 1949, that laid down in law a full set of protections, from working hours to health care. Even so, testimony from miners in the 20th century still provides a grim picture of the ponies' existence. Animals might be blinded when rubbing against sharp rocks, as they tried to find their way in the dark. On occasion, ponies panicked in a state of exhaustion, and in the subsequent struggle they, or even some miners, were injured or even killed. In one recorded incident in the Welsh coalfields, miners had to innovate to put down a badly injured horse, by pushing an explosive into its ear and firing the detonator. (See Ceri Thompson, *Harnessed: Colliery Horses in Wales*, 2008.)

The chief reason for the persistence of pit ponies in the age of mechanisation was their versatility. A pit pony could venture throughout most parts of a mine with relative ease, adjusting to variations in the ground and the height of roofs, turning easily and, of course, pulling heavy loads, not only of coal (although 85 per cent of coal was mechanically conveyed by 1952), but also any sort of load associated with mine maintenance and clearance. On the flip side, ponies were living creatures and, as we have already seen, could be prone to the natural behaviours of animals working in an environment for which they were never intended.

The pit pony truly lived life underground. It was winched down to the underground workings in a manner that must have been alarming for the animal: its legs bound tightly against its body, blindfolded, and positioned vertically (rump facing downwards) before being lowered the full depth of the shaft. Within the workings, there were stables with dedicated horsekeepers, responsible for the care and upkeep of the animals. The *Pit Pony* booklet explained that regulations required there had to be at least 'one horsekeeper to every fifteen ponies, but in practice there is generally one to every eight' (NCB 1952: 7). Full stabling was therefore provided underground, each pony having its own stall and generally proper illumination, at least in the age of electrification.

Once underground, the pit pony might spend the rest of its life there, without returning to the surface, although later they were brought up when miners got paid annual holidays – one week after 1938 and two weeks after 1948. Under the best legislative conditions in the 20th century, a pit pony was officially restricted to working no more than two 7½-hour shifts in a 24-hour

Pit ponies wore thick leather hoods to protect their heads and eyes from injuries resulting from scraping against the walls. *(Author/Big Pit NCM)*

An extant pit pony stables, the name of one of the horses still displayed upon the wall.
(Author/Big Pit NCM)

period, or three shifts in a 48-hour period. There was no fixed retirement age – the pony would keep working until it was no longer able to do so – although the average working life of the pit pony was 10–15 years old. Contrary to some myths, blind ponies were not worked down mines; for good reason, all animals needed to be fully sighted. The NCB booklet notes that when the animal was no longer able to work, the ideal was to move it back to the surface and rehouse it among the civilian population. It was recognised, however, that if no home could be found or there was a 'risk that he might fall into the hands of an unscrupulous dealer', the animal would be destroyed.

* * *

The percentage of tasks performed by manual labour in the coal industry, as we shall see in the next chapter, certainly decreased significantly from the late 19th century and during the 20th century. Yet coal mining remained a physically demanding job, regardless of what machines and technologies were there to supplement it. At the very basic level, even miners today need to understand how to read a coalface, wield a shovel and swing a pick.

A museum presentation of some of the basic tools of manual coal-getting.
(Author/Big Pit NCM)

Mechanisation: HAULAGE AND WINDING

In the 19th and the 20th centuries, coal mining underwent a series of technological revolutions in practice and productivity. The 287 million tons (292 million tonnes) of coal won in 1913 – the peak production year in the UK – would not have been possible without the invention and installation of new generations of machines that mechanised manual processes.

⬅ **A typical major haulage engine of the late 19th and early 20th century, the size of the winding drum indicating a deep mine.** *(Author/Big Pit NCM)*

The story of coal mining mechanisation is no less labyrinthine than any other aspect of the industry. In this chapter, we will look specifically at the major developments of the 19th and 20th centuries in coal haulage, exploring other aspects of mechanisation in the following chapters. Haulage – the movement of coal between the coalface and the pit surface – in itself might appear a rather niche focus. As we shall see, however, the efficient functioning of this process had an equal, if not greater, importance in relation to the actual cutting of coal. First, however, an overview of the core technologies behind mechanisation in general is essential to orienting these developments.

THE TECHNOLOGIES OF MECHANISATION

The machinery and power supplies that came to transform mining in particular were those that also reshaped industry in general. We can divide these into five categories:

1 steam power
2 compressed air
3 electrification
4 internal combustion engine
5 hydraulic power

STEAM POWER

Although experiments in using steam pressure to power machinery have an ancient heritage, the first practical applications came in the 18th century, in effect beginning the Industrial Revolution. We should also note that this century was the one in which the demand for coal leapt exponentially. This was propelled especially by Abraham Darby's first use of coke – produced by heating coal in the absence of air – rather than charcoal, to smelt iron in 1703; steadily, coke would come to be the fuel of choice for iron smelting, further elevating the volumes of coal required to keep Britain's industrialisation going. Coal-powered steam engines first appeared in mining in the 18th century, led initially by the need to find more efficient methods of implementing underground drainage. The first fully functional engine was that developed by inventor and engineer Thomas Savery in 1698, but although the high-pressure machine was labelled 'the miner's friend', its power output was too weak to perform the pumping job efficiently. Then, in the first decade of the 18th century, inventor Thomas Newcomen radically reworked the industrial steam engine. He used low-pressure steam to drive up a piston, then, at the top of the piston's stroke, the steam would be condensed with a jet of cold water, this creating a vacuum

Steam locomotives began to work in Britain's coal mines in the 19th century, primarily on the surface for hauling coal and waste. *(Author/ Big Pit NCM)*

beneath the piston. Atmospheric pressure would then drive the piston down to deliver its power stroke, at a cyclic rate of up to 12 strokes per minute, although later inventors such as John Smeaton further improved the speed and the power of the engine.

The main application of Newcomen's 'atmospheric engine' was to water pumping – whereby the piston was connected to the pump by a beam arm – but in the 1760s came the evolutionary step that gave steam wider practicality within the mining industry. James Watt, a Scottish mechanical engineer and chemist, transformed the efficiency of the atmospheric engine by fitting it with a separate condenser, which saved considerable amounts of energy being lost in the constant cooling process induced in single-cylinder designs. (Newcomen's engine had a single cylinder for both the injection and condensation of the steam.) Watt teamed up with the manufacturer Matthew Boulton, and working together they gave his engine further enhancements, including the use of low-pressure steam to assist the power of the working stroke. In the 1780s, orders for the new Boulton and Watt engine started to mount up in coal mining and other industries. Again, coal-mining usage was concentrated upon water pumping. But further refinements came once again in the 1770s and 1780s, such as using a sun and planet gear so that the engine could drive a wheel, introducing a double-acting steam system that made the machinery deliver smoother power outputs, and the centrifugal governor, which kept a constant speed under variable loads. Now the Boulton and Watt engine could be used not only for pumping in the coal mines, but also for use in winding machinery.

The next major figure in the evolution of British steam power was Richard Trevithick, an engineer from Cornwall. He looked at utilising high-pressure steam as the principal motive force for the engine, rather than atmospheric pressure, producing machines that created pressures of 145psi (10 bar); the other steam engines available at the time generated about 50psi (3.5 bar) at the top range of their power. These new engines were also far more compact than the others on the market, and thus were eminently suited to winding applications in mines. Trevithick further improved the power output of the Boulton and Watt engines, as well as developing the 'Cornish boiler' in 1812, which was stronger and more efficient than other types.

The steam engines of the landmark figures highlighted above, plus others who stood on their shoulders later in the century, changed the face of coal mining. Steam power brought not only energetic means of heavy winding, haulage (the first steam-powered haulage engines were installed underground in 1841) and pumping, but with the concomitant development of the railway network it transformed the system by which coal could be transported from producer to consumer. There were attempts to channel steam pressure beneath the ground via pipework to power other sorts of machines, such as mechanical cutters, but this was less than successful owing to the loss of pressure as steam condensed into water in the pipework.

A 'water-balance' headgear. A cage was at each end of the winding chain, and beneath each cage was a water tank that could be emptied or filled; the weight difference between the two tanks raised or lowered the cages. *(Author/Big Pit NCM)*

⬆ A compressed-air tank, used to power various pieces of machinery on the surface of the mine. *(Author/Big Pit NCM)*

⬇ An underground electrically powered rope haulage mechanism. This appears to be a main-and-tail system. *(Author/Big Pit NCM)*

COMPRESSED AIR

Between 1830 and 1850, ingenious minds saw other ways to convert the energy produced by steam into useful work at the coalface itself. One was the invention of the compressed-air engines, in which pressurised steam was used to compress air in a storage unit, the air then being fed out through pipes to work machinery.

Compressed-air systems had broad applications across industry, but it was particularly useful in mining for powering coal-cutting machines, which began to appear in volume in the 1850s and 1860s and continued in use throughout the 19th and 20th centuries. Compressed-air-powered locomotives also came into service, providing a new motive force for haulage. The compressed air was fed down to the machinery through either fixed cast-iron or flexible rubber piping, with connection points at regular stages along the main feed pipes. It was an ideal system for mining, as the compressed air did not involve underground ignition, plus no noxious gases were generated; in fact, the venting of the air into the atmosphere as the machinery worked actually helped with the ventilation of the tunnels.

ELECTRIFICATION

While steam and compressed air enabled the coal mining industry to shift up many gears in terms of its productivity, it was electrification that enabled it to go at top speed. The use of electricity has some safety limitations in mining, not least the importance of avoiding sparking and fire. But if these issues could be safely negotiated then electricity offered consistent, stable power for coal-cutting, haulage, lighting, heating, drainage, ventilation – indeed, any aspect of the mining process that required energy for work. The power could be supplied either through portable batteries (rechargeable and single-use), on-site generators and, eventually, via the National Grid. Furthermore, the machines driven by electricity reached levels of raw power that could not be matched by even the most impressive engines of the steam age.

In its adoption of electrical supply and equipment, Britain's coal industry actually lagged behind that of Continental Europe,

A horizontal winding engine from the early 20th century, manufactured by Walker Brothers of Wigan, a mining engineer company founded in 1866. *(Author/Boulton)*

ELECTRIFICATION IN BRITISH MINES

An article in *Nature* magazine from 3 December 1938 summarised key information given in a lecture about electrification in coal mines by a Mr R. Nelson to the Institution of Electrical Engineers, and in so doing provides an invaluable insight into both the history of mining electrification and the key debates about its use. The central points from the lecture were:

- In the year 1883, the first electric motor pump, rated at 15hp, was used to pump water from a coal mine.
- Britain's gravest colliery disaster – the explosion at Senghenydd Colliery, South Wales, in October 1913, which killed 439 people (a rescuer also died, to take the toll to 440) – led to miners calling for the removal of electricity from the pits. (The explosion was likely triggered when a build-up of methane was ignited by an electric spark from electric bell signalling gear.) This was avoided, however, by the introduction of stone-dusting measures (see Chapter 5).
- In 1937, a total of more than 2 million hp of motors were installed at collieries, half of them below ground.
- Accident statistics were revealing. It was recorded that between 1927 and 1936 electricity was responsible for 224 out of 8,656 deaths, or 2.5 per cent of the total loss of life in the pits.
- Electricity and compressed air were the main rivals in coal-cutting technology, although by 1937 a total of 70 per cent of the machine-cut coal was cut by electricity.

By the mid-20th century, manriding trains had been introduced to bring miners to and from the coalfaces. *(Author/Big Pit NCM)*

where we see electrical winding machines and coal-cutters in use from the 1880s. Slowly, the British began to catch up, so that by the end of the 19th century innovations such as portable electric lamps and electrical shaft lighting were becoming more common. It was during the first two decades of the 20th century, however, that Britain's mines more fully embraced the electrical revolution.

INTERNAL COMBUSTION ENGINE

The internal combustion engine, after a long phase of experimentation and development, reached a state of practical efficiency and reliability by the 1890s, mainly courtesy of a pantheon of German engineers and inventors, including Nikolaus Otto, Gottlieb Daimler, Wilhelm Maybach, Karl Benz and Rudolf Diesel.

The internal combustion engine is, in many ways, a poor fit with underground mining. The core mechanism by which such engines work – igniting highly flammable fuels in a continual stream of miniature cylinder-contained explosions – plus their toxic emissions, particularly carbon monoxide and carbon dioxide, make their applications in confined, explosion-prone environments problematic. (The use of diesel-powered generators and vehicles at the surface was, of course, straightforward.) Yet in the 20th century, improvements in engine design and mine ventilation meant that diesel power in particular could be utilised, mainly in the form of locomotives for haulage, but also for some power tools. The first such locomotives were introduced into the underground mines in the 1930s; the first diesel-powered locomotive with a 'flameproof' engine (i.e. one that prevents heat and flame coming into contact with the atmosphere) entering service in 1939. From that point on, the use of locomotives in mines leapt significantly, from 90 in 1947 to 906 just 10 years later.

HYDRAULIC POWER

The final critical ingredient in the mechanisation of British mines was hydraulic power, which utilised the incompressibility of liquids to generate work. Hydraulic machines first appeared in British mines in the second half of the 19th century, applied in water pumps (based on the principle that a small volume of water under great pressure can force a larger body of water under low pressure), and in drills and coal-cutting machines. Indeed, many of the modern electrically powered coal-cutters of mechanised mining (discussed in the following chapter) were in reality hybrid electro-hydraulic creatures, the electrical motor driving a hydraulic pump to actuate the cutter. Hydraulic power was also critical to the historical development of roof support mechanisation, with the first hydraulic roof support installed in British mines in 1947, the early hand-operated types later evolving into the computer-controlled banks of self-advancing supports (see Chapter 5).

* * *

We have now established the key energy sources underpinning mechanisation in the British coal industry, and their main applications. Now we will look more closely at the specific types of machinery involved, beginning our journey with that most critical activity: moving coal from the coalface to the surface.

TYPES OF TRANSPORT

A useful way to unpack the mechanisation of coal haulage is to look at how the haulage systems of an underground mine operated in the 1950s. By this era, most of the key advances in mechanised haulage had already taken place, but many of the core principles used in 19th-century mining still applied, just with different power sources (electricity and diesel instead of steam). We will make historical digressions and explanations as required.

Coal mine transport was broken down into three key stages:

1 **Face transport** – This refers to the haulage of the coal from the point at which it is cut, moving it across the coalface to the gate (the access tunnel leading to the main roadway).
2 **Secondary transport** – This relates to the conveyance of the coal from the end of the face, then along the gate to the point

Powered coal scoops mechanised some of the back-breaking coal loading and shifting work previously performed by hand. *(Author/Big Pit NCM)*

➡ **A face conveyor belt transfers its load of freshly cut coal to a gate conveyor at a transfer point.** *(Author/ John Cornwell/ Big Pit NCM)*

⬇ **A loader chute at the end of a conveyor belt delivers its load into waiting coal tubs.** *(Author/Mason/Richard Sutcliffe Ltd)*

at which the secondary transport connects with the main transport system.

3 **Main transport** – The coal haulage system that runs directly to the shaft along the mine's main roadways.

Each stage of the journey classically required a different method of coal conveyancing or haulage. The transition points between these stages of the haulage system might be relatively seamless, but at some point a loading station would be required: a point at which the coal is transferred physically or mechanically between two different forms of transport (e.g. from a belt to a locomotive-drawn tub or car). Because loading stations essentially represent nodes at which the haulage system slows down, official NCB recommendation was to keep them to a minimum; it was preferable to have a single large and efficient loading station than multiple inefficient ones. Loading stations in well-equipped mines were indeed advanced; a single human operator could take charge of the loading station machinery, with loader chutes directing the coal into tubs or on to other conveyors. (The loader chutes were important to control the drop of the coal, to avoid smashing the large pieces into less useful smaller pieces.) The three stages of transport outlined above apply most clearly to longwall mining, as the distinction between face transport and secondary transport is not so clear in pillar-and-stall workings.

FACE TRANSPORT

Prior to 1900, the coal cut from the face was loaded into corves or tubs directly. During the early 20th century, however, mechanisation saw the steady switch to electrically powered conveyors at the face. By the mid-1950s, more than four-fifths of coal output was carried by face conveyors, while only about one-sixth was hand conveyed into tubs.

The basic conveyor was a belt, running around a driving drum at one end and a loop drum at the other, the belt carrying the coal to the point of delivery into a tub. This simple explanation, however, expands under further analysis, as the belts themselves and the mechanisms that drove them could vary.

Three basic types of face conveyor were used:

1. **Belt conveyor** – This was a wide belt running continuously between the face and the gate, although depending on the pit, belt conveyors might also be used in the secondary and main stages of transportation as well. The advantages of the belts were that they were suitable for use on moderate gradients (up or down), were mechanically simple and were relatively straightforward to install and to move up to the coalface after each cut.
2. **Scraper-chain conveyor** – This conveyor featured a trough with powered chains running within it, the chains sectioned by cross-arms that drove the coal onwards. The chief advantage of the scraper-chain conveyor was that it worked better than a belt on steep gradients, the cross-arms preventing the coal from rolling off. It could also be used slewed; this enabled the

Miners pose for a photograph next to filled mine tubs, on what is probably a main transport route. *(Author/Big Pit NCM)*

CARS, TUBS AND SKIPS

The increasing power of underground locomotives and rope haulage systems during the 20th century contributed to the evolution of the 'mine car'. This was little more than a large four-wheeled tub, but its carrying capability was dramatically expanded. Each tub carried half a ton (0.5 tonnes) of coal, thus eight such tubs arranged in a train could transport 4 tons (4.06 tonnes) of coal in a 32-wheel chain of vehicles. A single coal car, however, could pull 4 tons alone, with just four wheels in contact with the rails. The NCB handbook *Moving Coal Underground* therefore observed that 'To move and load coal with mine cars instead of tubs means fewer cars, fewer men, fewer wheels, less friction, less wear and tear, less maintenance, less driving power, better constructions, higher speed, less breakage of coal, less coal dust' (NCB 1957: 13).

In later years, very large receptacles called skips were used for hauling the coal to the surface. A skip could carry up to 20 tons (20.3 tonnes) of load, depending on its specifications, and was filled at the pit bottom directly from a haulage conveyor or a feed from an underground coal bunker. Once the skip reached the surface, the coal was then unloaded, often automatically, via the skip's integral chute. Skips certainly multiplied the volume of coal that could be moved in one go, but their use was not just a list of advantages. They tended to produce a high volume of coal dust, for example, and large pieces of coal were often broken down into smaller pieces during the loading and unloading processes.

This photograph usefully shows a comparison between the smaller coal tubs and a larger 4-ton coal car at the rear. *(Author/Big Pit NCM)*

belt to be pushed up to the working face behind the cutter as the cutter advanced. The 1956 *Deputy's Manual* also noted that scraper-chains 'are most suitable for gate-end loaders and have a wide application as the link between the face belt and the gate belt, in the role or stage loader or feeder. This is because chain scrapers will deal with an eccentric load and can be arranged to deliver it uniformly in line and in the proper place on a belt' (Mason 1956: 472).

The scraper-chain type of conveyor evolved into what the NCB termed the Armoured Flexible Conveyor (AFC), which, as we shall see, was typically part of a combined cutter-loader system. Unlike the belt conveyors, with their flexible synthetic belt, the AFC was constructed of heavy-duty steel; indeed, such machines were often referred to as 'Panzer' conveyors, using the German word for 'armour'

3 **Shaker conveyer** – This type of conveyor did not have an endless chain or belt. Instead, they utilised forward motion interrupted by constant sudden short backward strokes that caused the load, under its own inertia, to be shaken forwards in the required direction. Shaker conveyors were often married to a duckbill loader, which was a broad shovel-like fitting on the end of the conveyor; this fitting shook with the conveyor and thereby drew the cut coal up on to the belt. These mechanisms were best used on downhill inclines of 1:4.

SECONDARY TRANSPORT

The secondary transport (along the gate to the main roadway) in an underground mine was performed either by new lengths of the conveyor system or, if the coal had already been loaded into coal cars or tubs, pulled by locomotives. Gate conveyors tended to have wider belts and travelled longer distances than the face conveyors, and it might be that a single gate conveyor would be receiving inputs from several faces at various points along its length. The gate belt would typically run into a loading station, at which point the coal would be loaded into cars or tubs, which were in turn hooked up to either locomotive or rope haulage systems for the final journey up the main transport roadway to the shaft.

A scraper-chain conveyor, well suited to taking the coal up inclines with limited spillage. *(Author/Big Pit NCM)*

Underground locomotive haulage actually began back in the 1840s. The fire and heat of the steam engines meant that these vehicles could only be used in areas of the mine that were free from firedamp, but they provided an alternative to horse-drawn motive power. The train ran on narrow-gauge lines, warranted by the width of the tunnels. Common gauges were: 1ft 10in (0.56m), 2ft (0.61m), 2ft 6in (0.76m) and 3ft (0.91m).

Compressed-air locomotives

Mining locomotives designed to run on compressed air entered service in the 1870s. Compressed-air locomotives, by their very nature, were far safer to use underground than their steam predecessors, their power source being large tanks carrying the pressurised gas rather than the flames and boilers of steam engines. Here lay a problem, however. The compressed-air cylinders were very quickly depleted when pulling heavy loads, especially in the early versions of the engines, which

➡ **An electro locomotive transports machinery to and from the coalface. This appears to be a battery type rather than a trolley-wire type.** *(Author/Big Pit NCM)*

⬇ **Diesel-powered tractors with rubber wheels became heavily used on the main roadways of mines. Note also here the roof support system, a combination of steel arch girders, steel struts and ties and corrugated iron sheets.** *(Author/Big Pit NCM)*

by the late 1870s had a maximum pressure of 100–200psi (7–14 bar). This meant that at best the locomotive might only be able to pull the load several hundred feet before a recharge was necessary. Thankfully for the efficiency of the system, recharging was relatively fast and straightforward. Recharge stations were dotted throughout the mine, these being fed by compressors on the pit surface. Gradually, the cylinder pressures of these locomotives were improved, so that during the first decade of the 20th century they had reached about 800psi (55 bar) for single-tank models, and 1,000psi (69 bar) for double-tank engines.

Electric locomotives

Compressed-air locomotive haulage quickly peaked in terms of its efficiency in the early 20th century, and although we see such engines continuing to work in various mines well into the 1900s, they were progressively replaced by electric or diesel locomotives. Electric locomotives came in two basic types: 1) those powered by a pantograph and trolley wire system; 2) those powered by battery storage. Given the risk of sparking from the contact between the pantograph and the power supply, the former type was only permitted to be used in a 'non-safety lamp mine', i.e. one in which there was no threat of firedamp. The voltage was kept relatively low for safety, *c*. 250 volts. This type of locomotive relied upon the installation of a power wire, properly tensioned, along the roof of the transit tunnel. Stipulations stated that this wire had to be a minimum of 7ft (2.1m) in height from the rails and at least 1ft (0.3m) from the roof and the sides. If the locomotive had to travel into parts of the mine that were not suitable to the installation of the overhead wire, then it could adopt a cable-reel mechanism. Here, an electrical cable, wound on a drum at the back of the locomotive, was connected to the power supply and, as the train travelled forwards, the cable unwound and was laid behind the locomotive on the track. On the locomotive's return journey, the power cable was wound back on to the reel.

Electric storage-battery locomotives eventually replaced the need for cable-reel power, and the performance of these types became effective enough to offer comparable performance to diesel types. In fact, as the *Deputy's Manual* noted, electric-storage battery locomotives could be 'favoured' over diesel in deep mines because 'there are no hot gases to increase further the already high ventilation temperatures, no noxious gases are fed into the intakes, there is no smell to mask the detection of spontaneous heating, and much less heat is developed in the motors than in the diesel engine' (Mason 1956: 448). Mason goes on to point out a further advantage of such vehicles, in that they could be maintained by the colliery's electrical staff; they didn't need automotive engineers, unlike the diesel vehicles.

The main extra logistical requirement of the electric storage-battery locomotives was a battery-charging station. Given the risk of fire or sparking, there were very strict regulations governing where the charging station was located and how it was maintained. All flammable materials had to be removed and smoking was strictly forbidden. The station also had to be well ventilated, with a through current of air passing over the apparatus.

Diesel locomotives

Diesel locomotives became the most prolific of the types working in British mines. This was partly because of their performance – they could pull loads up gradients often not accessible to the electric vehicles – but also for their convenience, as they did not require the laying of overhead cables. Furthermore, while petrol engines were banned from use in underground mines, because of their electrical spark ignition systems, diesel engines were permitted since they work by raising the temperature of the air in the cylinder due to mechanical compression (adiabatic compression), sufficient to ignite the fuel.

Nevertheless, diesel locomotives did require some special adaptation before they could work the mines. Particular attention was paid to the exhaust system, which was modified to prevent emissions of both flame and poisonous gases. The flame trap and conditioning system worked in several stages. The exhaust gases were sprayed with water from a fitted water tank and sprayer system to cool them down, an action that also trapped the soluble gases. The flame traps were formed from thin plates of high-quality stainless steel, separated by fractions of an inch; by passing through these gaps, potential flames were nullified. Even with these measures, diesel locomotives still emitted the particularly dangerous non-soluble gases carbon dioxide and carbon monoxide, and there were strict stipulations that diesel locomotives could not be used in air that contained 0.1 per cent carbon monoxide.

In addition to the locomotives used for pulling tubs and cars, another coal transporter that might be used on secondary roads was the shuttle car. This vehicle was, in essence, a large four-wheeled truck that carried the coal within the body of the vehicle itself. The utility of the shuttle car was that it did not run on rails and had more manoeuvrability within the mine, hence it was generally used in pillar-and-stall workings. The shuttle car was designed so that its load could be deposited directly into or on to the next form of transport.

⬆ Coal is transported towards the pit bottom on a standard belt conveyor. The track to the left is for bringing supply drams (tubs) onto the coal face. *(Author/Big Pit NCM)*

➡ A shot of Maltby Colliery, South Yorkshire, showing the above-ground conveyors (in metal housing) linked to No. 3 Main Shaft. *(Alan Murray Rust/ Conveyors at Maltby Colliery/CC BY-SA 2.0)*

MAIN TRANSPORT

Once the coal had reached the main roadway it could now begin the final part of its underground journey, up to the shaft bottom. By the 1970s and 1980s, this part of the journey was often fulfilled by additional conveyors, but also by the locomotives, although there were definite limitations to this latter method. With some variation according to the locomotive type, the rail configuration and the load pulled, a gradient of 1:12 was generally the limit of practical haulage by locomotive, although this sometimes could be exceeded through the expedient of scattering sand on the rails to improve traction. In the days of horse-drawn haulage, pit ponies and strong draught horses could pull loads effectively up more precipitous inclines, but their rate of work could not be sustained over long periods of time, as opposed to mechanical means.

During the 19th century, therefore, the principal system of main transport haulage came to be a rope or chain mechanism powered by an underground winding engine. A powerful winding engine could draw the heaviest loads of coal up almost any manner of gradient; the principle was in essence an extension of the vertical winding used in shaft haulage and movement.

The first underground steam haulage engines were introduced in the early 1800s. They did the job required of them with reasonable efficiency, but the main problem was venting the smoke produced by their boilers. This was functionally achieved by a long flue that carried the smoke out to the surface, but the proliferation of underground winding engines still fostered a thick haze of smoke in the tunnels. During the second half of the 19th century, therefore, an effort was made to move the boilers to the pit surface, transferring the steam down to the haulage engines via pipework, although as noted, the condensation of steam en route led to a reduction in power.

In 1850, the first compressed-air winding engine came into service, in a colliery near Glasgow, and gradually these machines replaced steam-driven rope haulage engines during the remainder of the century. In fact, even with the introduction of electrically powered rope haulage mechanisms, from 1883 – which gave new levels of efficiency and power output – compressed-air types remained in service well into the 20th century. (*Moving Coal Underground* stated that about one-third of rope haulage engines were still driven by compressed air at the time of publication, the mid-1950s.)

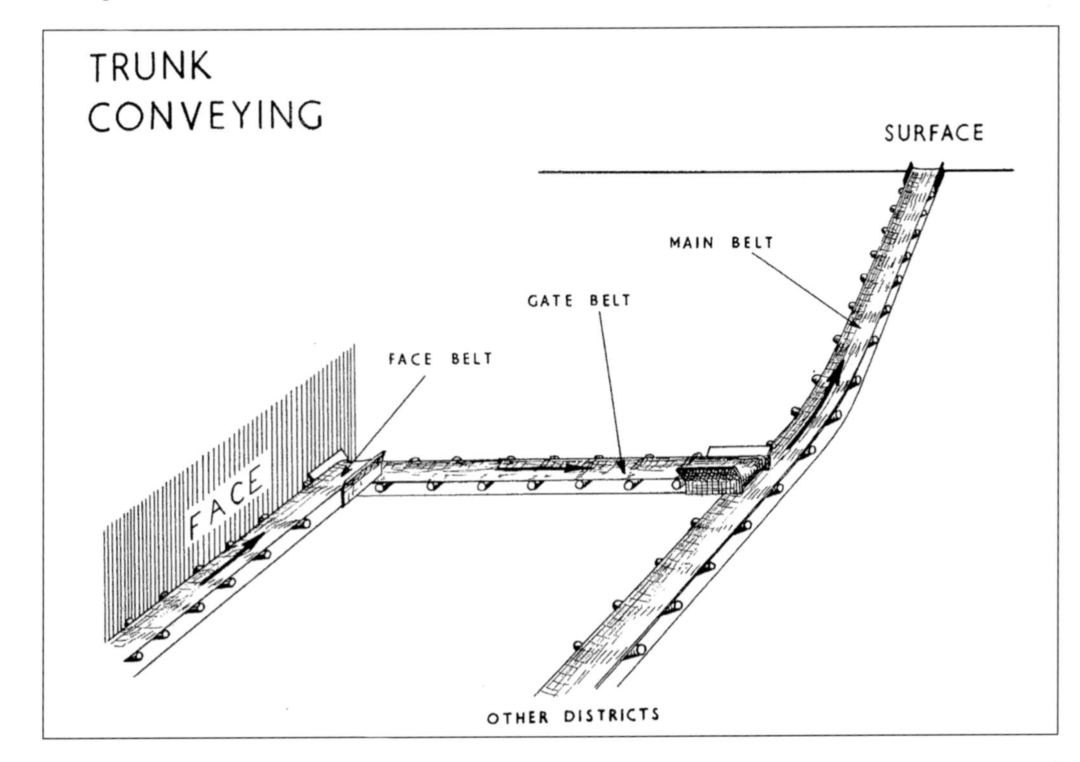

A simplified National Coal Board diagram showing the sequence of conveyors that took coal from the face to the surface. *(NCB)*

An elevation and plan diagram of an electrically driven winding engine, manufactured by the Harpener Mining Company, Dortmund, Germany. *(Author/ Boulton)*

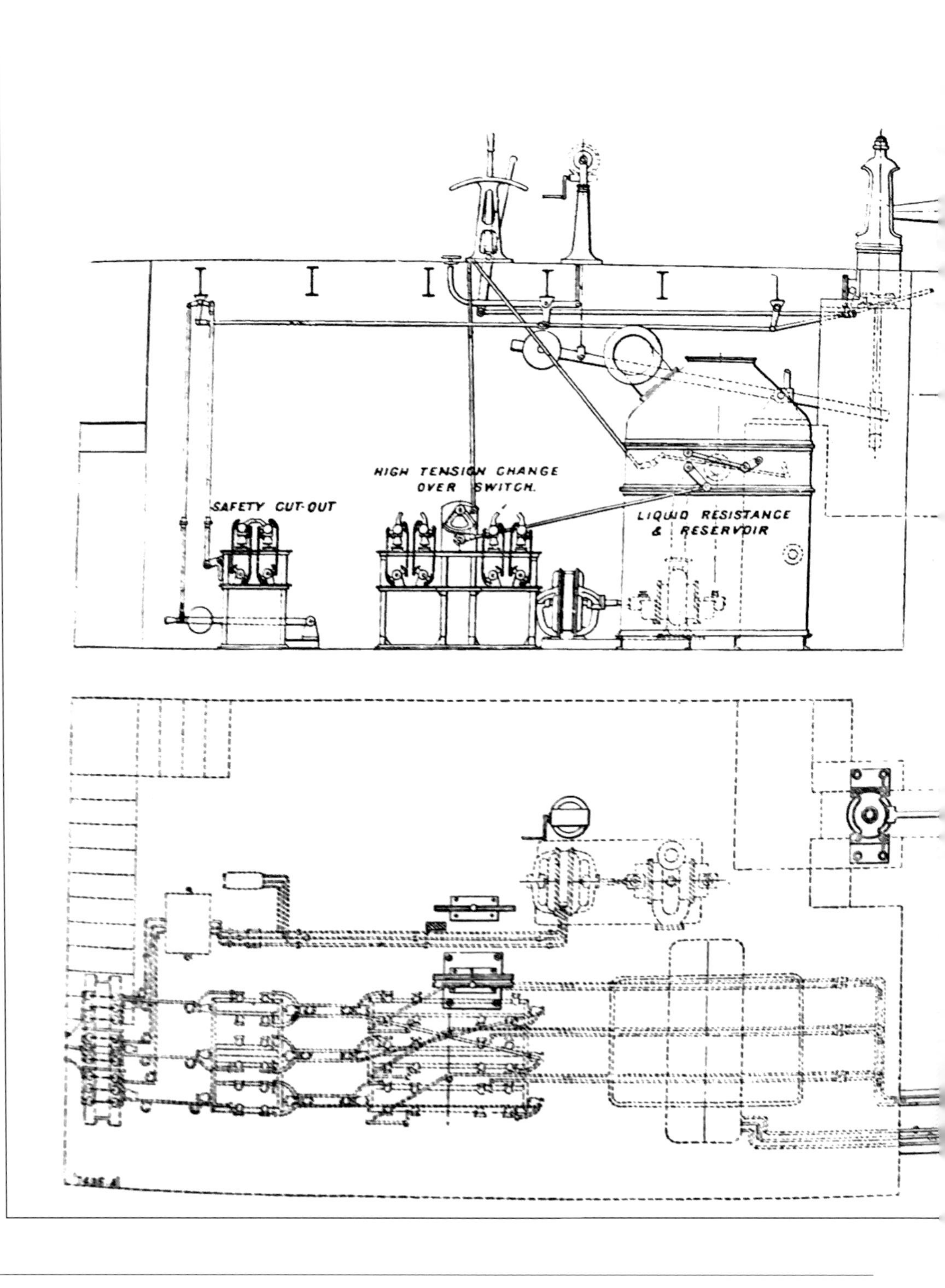

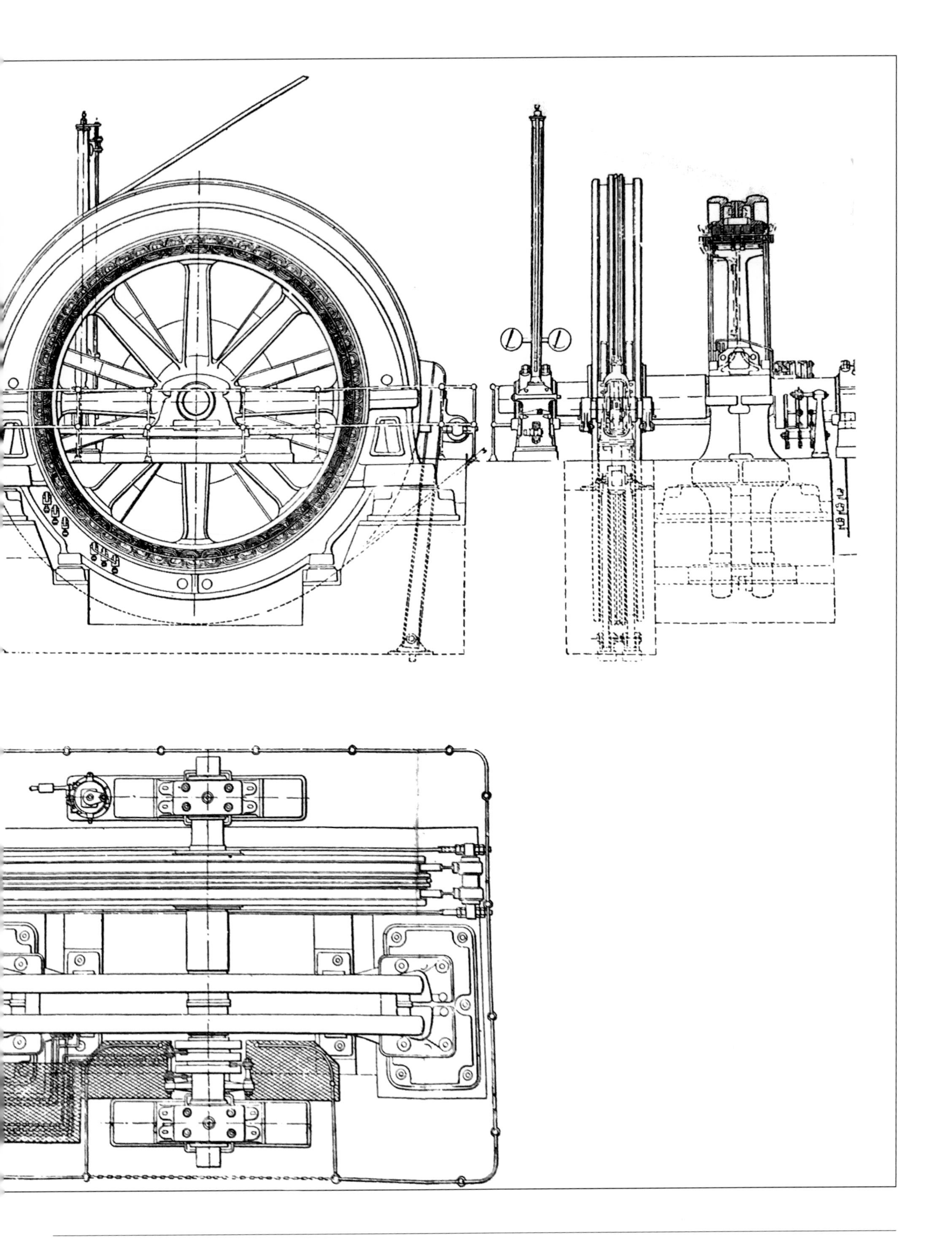

WINDING ROPES

The physical design of wire ropes has a particular importance in the context of coal mining, with the different types of rope having contrasting characteristics, not just in terms of breaking strength, but also relating to their diametrical volume (which affects the depths at which they spool up on a winding drum), flexibility and durability. The two main categories of rope are stranded ropes and locked coil ropes.

Ordinary lay stranded ropes are the most basic type, with a single central wire around which the outer wires are twisted. This type of wire has the advantage of flexibility, but it is prone to wearing and abrasion, so in coal mining it was generally limited to use on small haulage machines, cranes, scaffolds and cutters, rather than heavy haulage and winding engines.

Lang's lay ropes (another stranded type), however, undergo extra twisting during manufacture and the outer strands are flattened, making the wire less flexible but far stronger, and thus usable for heavy haulage and winding.

Multi-stranded, or compound stranded, wire tackles the problem of stranded wires twisting and untwisting under variable loads, progressing further strands of rope around a smaller complete rope, but running in the opposite direction to perform a locking effect. These wires were principally used for Koepe friction winders (see below) and for tail ropes.

Locked-coil ropes feature interlocked Z-section wires on the outer part of the rope, protecting the inner core of the rope from unravelling and from environmental damage, although they also have reduced flexibility. They are most commonly used for winding.

Mining engineer Matthew Saunders notes that 'The selection of a winding rope depends upon several factors, not limited to: a) Dynamic Working Load + factor of safety; b) shaft depth; c) shaft conditions (temperature/water/humidity); d) type of winding system. The effective spooling diameter on the drum is also important, as the rope may only be bent 80–120 times its diameter dependent upon design. Often winding systems are specified as complete installations and winder type and rope type are a harmonious decision taking into account factors from each.'

Sections of various types of wire ropes typical of those used in coal mining for winding and haulage. *(Author/Big Pit NCM)*

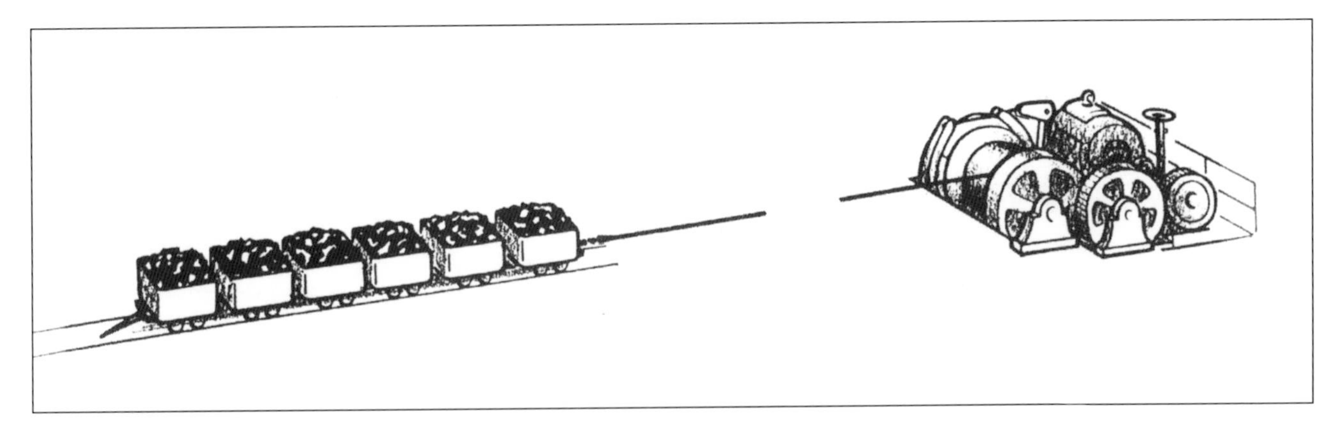

A simple single-wire direct rope haulage arrangement, the tubs running on a single track. *(NCB)*

ROPE HAULAGE

There were several options for how a rope haulage system could be configured:

- direct (or main) rope
- main-and-tail rope
- endless rope
- self-acting rope

DIRECT-ROPE HAULAGE

Direct-rope, or main-rope, haulage was the simplest of the methods, with a haulage engine featuring a single winding drum and a single rope, these used both for pulling the loaded cars or tubs up the roadway and lowering them back down again. The physics of this system demanded that the roadway be on an incline outbye, as the empty cars would travel back down the road under gravity assistance, the winding drum restraining the descent through engine or brake resistance.

Direct-rope haulage could also be subdivided into single-track or double-track methods. The single-track method is essentially that described in the previous paragraph, with the cars or tubs ascending or descending on the same single-track rails. In double-track haulage, there were two tracks running side by side, one for ascent and one for descent. The haulage mechanism was configured so that one end of the rope would pull the full cars up while the other end of the row controlled the downward movement of the empty cars.

ENDLESS-ROPE HAULAGE

By the 1950s and 1960s, endless rope was the most common of the rope-haulage systems in use in British mines. The endless-rope system consisted of two parallel tracks, one for full tubs moving outbye and one for empty tubs moving inbye. Both sets of tubs, however, were attached to the same rope, which passed around a powered driving wheel on the engine inbye (a motor powered a surge wheel around which the rope passed several times) and a return-wheel mechanism at the outbye end. The system had to be kept under tension to prevent the rope slipping, especially as unequal loads would be pulled on the different tracks at the same time. One or both of the rope wheels at the end of the mechanism were placed under tension, with a tension rope connecting to the wheel housing, running up over pulleys, and terminating in a large tensioning weight that dangled down into a pit. The tubs and cars connected to the

A close-up of a scraper-chain conveyor. The scraper arms are bolted directly to the driving chain. *(Author/Big Pit NCM)*

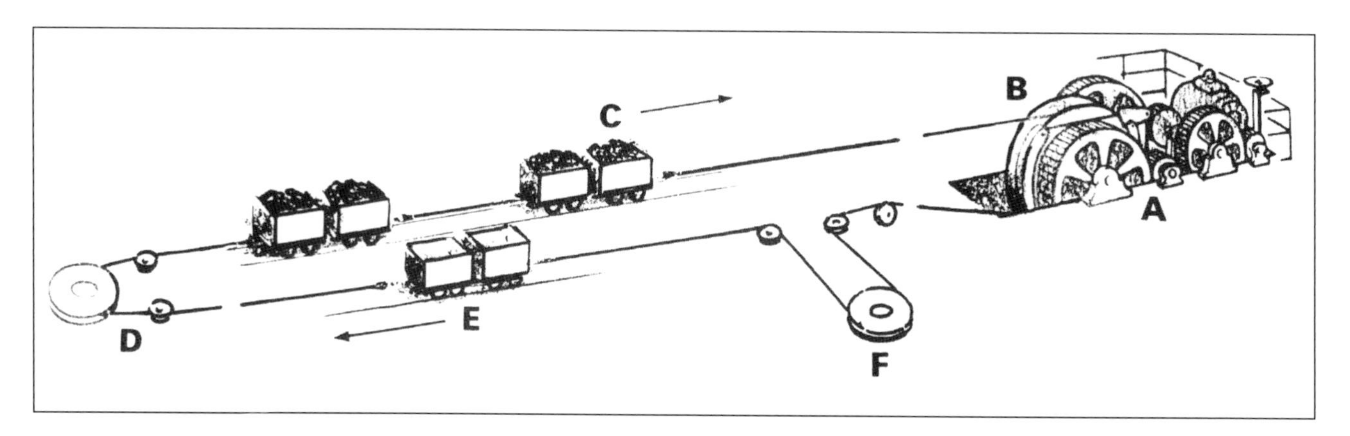

Endless-rope haulage
A Haulage engine
B Surge wheel (or pulley)
C Full tubs
D Return pulley
E Empty tubs
F Rope tensioner
(NCB)

endless rope via special locking mechanisms that clamped directly on to the rope at designated points.

Endless-rope haulage was reliable and effective; some systems would run 10,000 tons (10,160 tonnes) of coal along 2 miles (3.2km) of roadway every single week. The system was only optimal, however, on long and straight roadways without significant deviations. Former miners also remember that the clamp mechanisms on the tubs could be hard on the ropes, squashing the wires and causing rope stretch. A temporary (and unauthorised) expedient to overcome rope slipping was to wrap a piece of cement bag around the wire to increase the wire diameter and grip, and fix the clamp around this.

MAIN-AND-TAIL HAULAGE

This type of haulage offered a solution to several problems, particularly if the roadway did not have enough of a downward gradient inbye to use the direct-rope system, or if the roadway undulated up and down across its distance. It also could be used on roadways that did not have enough width for double parallel tracks, and which were therefore not suitable for the endless-rope mechanism.

The main-and-tail system utilised two ropes: a main rope and a tail rope, each connected to two separate winding drums set side by side at the engine end of the mechanism. Each winding drum has its own brake. The main rope was coupled to the front of a full load of tubs (or what was the back of an empty row of tubs). The tail rope, when the tubs were heading outbye, was linked to the back of the rear tub; it went backwards around a pulley at the far inbye end, then returned back on itself to the tail rope winding drum, running on a series of rollers to keep it clear of the moving tubs, the ground and any other traffic. As the full tubs headed outbye, the main rope wound on to its drum, while the tail rope unwound. When the full train eventually reached the winder end, the operator disconnected both of

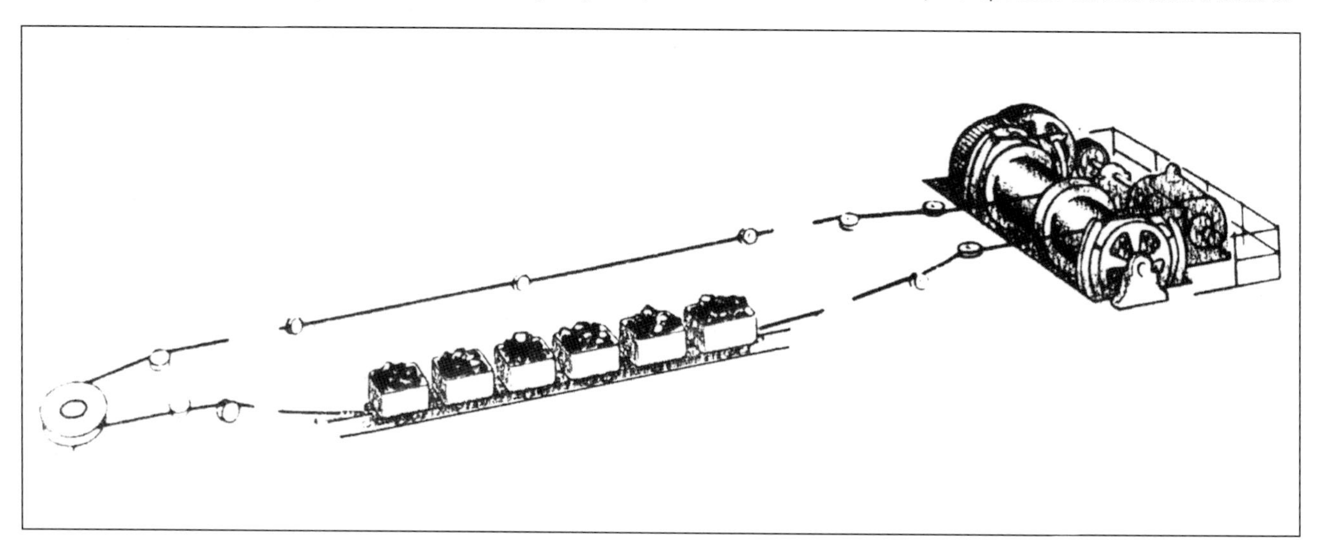

A main-and-tail haulage system, the main rope connected to the front and rear of the sequence of coal cars. *(NCB)*

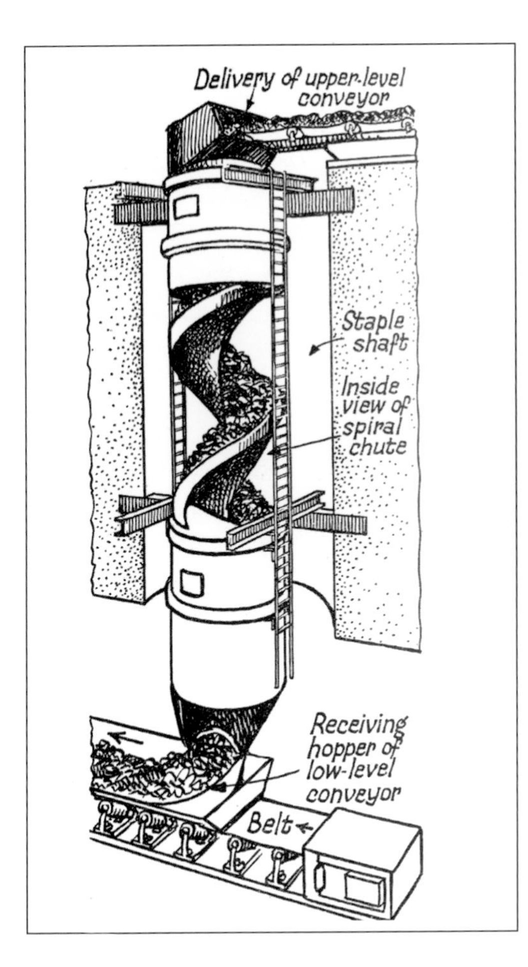

Transferring coal between split-level conveyors using a spiral transfer chute. *(Author/Mason)*

the ropes and attached them to an empty train. Now the tail rope was wound in on its drum, which pulled the empty tubs inbye, while the main rope was fed out; the main rope during this stage was protected by running through a set of rollers in the centre of the track.

SELF-ACTING ROPE HAULAGE

Self-acting haulage methods were actually among the earliest forms of haulage systems, in that they needed no external mechanical power – they were purely gravity operated, as long as the roadway gradient had an incline of 1:20 or more in favour of the load (i.e. inclined down towards the outbye). In the absence of a powered winder, the system simply used a pulley and brake system, the brake controlled by the person at the top of the incline.

Having described tub haulage systems in

MOVING MEN AND SUPPLIES

It was not only coal that required efficient transportation around the underground mine workings. Supplies and men also needed moving, often across distances measured in miles rather than metres. For transporting supplies, by the 1970s the monorail system was in use in many mines, the supply cars running on a monorail track attached to the roadway roof girders and operated either by endless-rope or main-and-tail haulage or pulled by diesel locomotives. This system had the advantage of travelling over, rather than across, uneven floors, but their weight and that of their loads could distort the roof supports. Another option was the coolie car, a flatbed vehicle that ran on conventional floor-mounted rails, with eight wheels carrying the car and a further eight gripping the track sides horizontally to prevent derailment. By contrast, the Becorit Roadrailer system ran on a single central rail, pulled by a diesel locomotive.

When it came to transporting men, in times past workers would often simply catch a ride on a passing conveyor, although this dangerous practice was eventually prohibited. In its place, different systems of 'manriding' were installed. The most typical arrangement was a train of seated carriages, called a manrider, paddy or spake, pulled by a diesel locomotive or rope haulage. For steep inclines, however, a ski lift-type system was used.

A manrider takes miners out to the coalface for their shift. This one is a rope haulage type. *(Author/Cornwell/Big Pit NCM)*

➡ **A rope haulage track (left) and a belt conveyor sit side by side in a roadway. Note how the belt is suspended above the ground, to facilitate cleaning.** *(Author/ Cornwell/Big Pit NCM)*

➡➡ **This diagram shows how the coal belt is supported by two parallel steel-wire cables, running over pulleys.** *(NCB)*

HAULAGE SAFETY DEVICES

A diagram showing the safety catch system for restraining a runaway tub. *(Author/ Boulton)*

The dangers of runaway cars and tubs were very real on long roadways with steep inclines; accident logs well into the 20th century show such runaways as a not-uncommon cause of death among miners. According to accident reports, many of these accidents could have been avoided by the correct implementation of all signalling procedures between the operators inbye and outbye, the careful and frequent inspection of haulage wires and ropes, and the proper laying and maintenance of tracks.

During the 20th century, however, a range of mechanical safety features were introduced to help guard against runaway accidents. The specific features were:

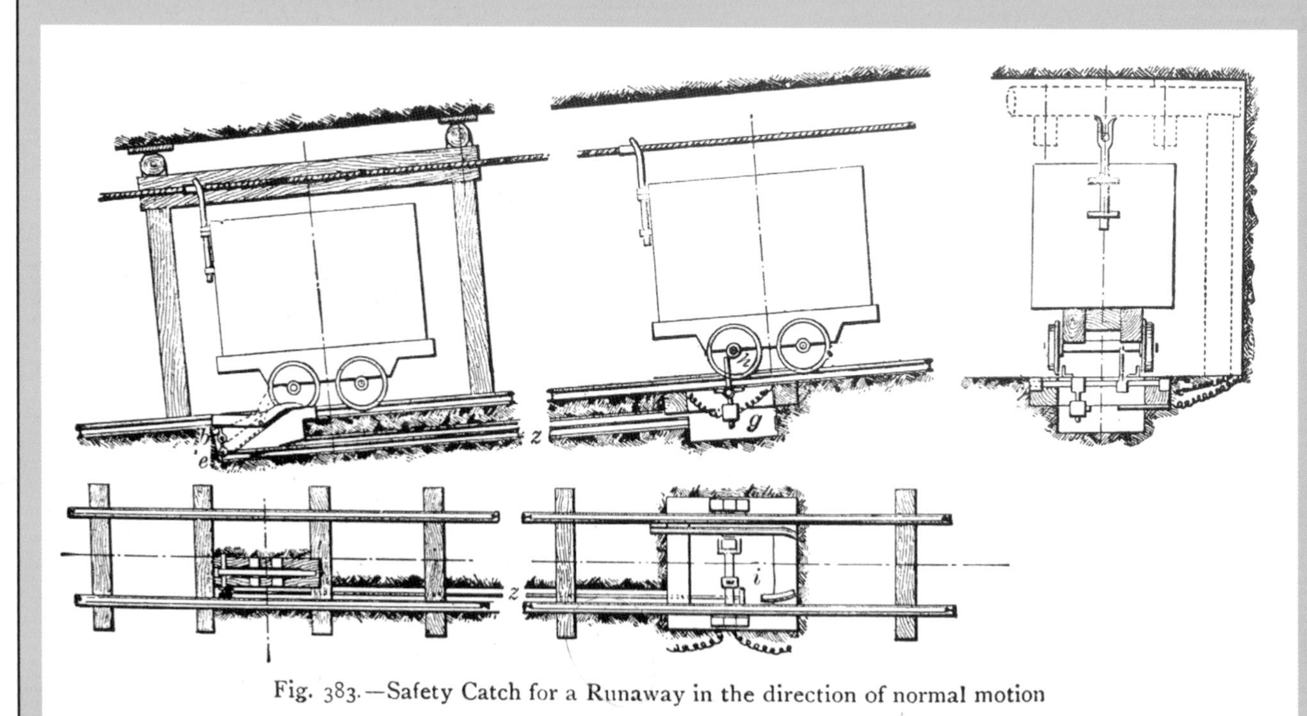

Fig. 383.—Safety Catch for a Runaway in the direction of normal motion

detail, we should also note that in many of the modern mines of the 20th century conveyors were often used to run the coal all the way up the roadways to the pit bottom. This process depended upon roadways that were long and straight, and maintained (ground movements could introduce unwelcome angles and undulations into the roadway). To avoid blocking transit of tunnels, conveyors could also be roof-mounted rather than set at floor level. In many ways, this was preferable for reasons of safety. Conveyors close to the floor could build up coal dust beneath, which posed a fire hazard, whereas coal dust falling from overhead conveyors would lie on the exposed floor, where it was more accessible for cleaning up.

- **Drop Warwick** – Long beams fitted to the roof girders directly over the tracks by a hinge at one end; the other end was held flush to the roof by an eye-bolt and pin. In the event of a runaway tub or car, the pin was pulled out from the eye-bolt by a wire, releasing one end of the beam, which dropped down on to the track to stop the moving tub. Most miners, however, recall that the girder was always hinged down to stop runaways and was only raised when a journey of tubs needed to come through.
- **Monkey catch/back catch** – Simple pivoting arms fitted at regular intervals on the sleepers between the rails. The placement of the pivot point caused the arm to stick up in the air facing the outbye direction, but were designed so that the axles of the full cars/tubs would travel over them without problems when heading up an incline. Should the loaded cars roll back downhill, however, they would be stopped by the raised arm.
- **Back stay** – Back stays were beams that were dragged behind the last car of a full load; if the tubs started to run away, the back stays would dig into the ground to arrest the movement.
- **Locker** – An iron bar placed between the spokes of the tub to lock the wheel into place.
- **Dumpling** – A wedge-shaped hardwood block placed between the haulage tracks to stop a runaway.
- **Scotch block** – A wedge block, fitted with a handle for use, that was placed in front of the wheel of a tub, rather like a wheel chock in aviation safety.
- **Stop block** – A block of metal-sheathed hardwood placed across or between the track to stop overrunning tubs.
- **Jazz rail** – A swerving section of track specifically designed to throw off speeding tubs.
- **Runaway switch** – A track switching mechanism used to derail the tub.
- **Retarder** – A mechanism fitted to tub wheels or axles, designed to reduce the excessive speed of a runaway tub or car.
- **Squeezer** – Converging and pivoted horizontal beams that grip the sides of the car and slow it down.
- **Manchester gate** – a horizontal girder hinged between steel posts; the girder was always closed unless journeys needed to pass through.

For safety, roadways would also feature refuge holes (or 'man holes') cut into the roadway sides at regular intervals, in which the miners could take shelter to avoid runaway tubs and cars.

⬆ **Coal cars sit on a double track system at pit bottom, awaiting filling and transfer to the surface.** *(Author/Big Pit NCM)*

⬇ **The headstock at England's National Coal Mining Museum (Caphouse Colliery).** *(Plucas58/CC0 1.0)*

WINDING GEAR

Without effective winding gear, there is no deep coal mine. Prior to the invention of steam power, as described previously, the only methods of winding coal, machines, material and men up and down the shaft were human- or horse-powered manual winding systems, with all their limitations of speed and power. (In some of the deepest early mines, the men and loads travelling along deep shafts might be compelled to make two or three stops at designated points while the winding machinery above was reset.)

Another option used in some collieries was the overshoot waterwheel, specifically for those pits located next to fast-flowing rivers. Waterwheel winding could deliver respectable performance, but the speed and viability of water winding depended very much on seasonal and environmental factors; a long, dry summer could cause water levels and current strengths to drop significantly, with a consequent impact on winding speeds and weights.

STEAM WINDING

In the 18th and 19th centuries, steam shaft winding arrived at the pithead – as it did elsewhere in British industry – the first examples appearing in the 1760s. One interesting early development was the 'water-coal gin', developed by an engineer from Yorkshire, John Smeaton. In this system, Smeaton actually used the water yielded by a steam-driven pump to drive waterwheel-powered winding equipment. The system was installed in a few collieries in the north-east of England. At those collieries that could afford the investment, however, steam engines of the Newcomen and Watt types, fitted with crankshafts and winding drums, became the most common winding types in the second half of the 18th century. These beam-type engines were not the most efficient winding tools, but they revolutionised coal mining, enabling heavier loads to be brought up from ever-greater depths.

A critical mechanical innovation came in 1800, when engineer Phileas Crowther developed a significant modification of the steam winding engine. In the 'vertical winding engine', the winding drum and flywheel cylinder were positioned above the steam cylinder, which drove an overhead crankshaft directly. The system was both faster and delivered more power, and it became well established in the collieries of the north-east. The presence of such an engine was often indicated by particularly tall stone- or brick-built engine houses, their elevation required to accommodate the tall machinery.

In the middle of the 19th century, however, horizontal high-pressure winding engines came to be dominant, and remained so until electrification took over completely. The horizontal winding engine had the cylinder, crankshaft and flywheel set on a horizontal plane, the flywheel rope running up from ground level to the cage sheave wheel set forwards and up on a headframe directly above the shaft. It was these headframes that became the familiar landmark of coal mines – and remain so to this day.

There was much highly technical debate about the relative virtues of vertical or horizontal engine configurations, involving finely tuned arguments over matters of leverage and power. These are beyond the scope of this book. Boulton, however, summarised the 'advantages of the horizontal engine' in quite practical terms: '(*a*) the engineman has the whole of his engine within sight, and can thus see that the ropes are coiling evenly on and off the drum, and that every part of the engine is working properly; (*b*) the foundations being more rigid, the engine works steadier; (*c*) the engine is easier of inspection and examination' (Boulton 1908: 86). Boulton could have added that the new generation of steam winding engines were also faster, producing speeds of up to 90ft/sec (27m/sec) in some of the deepest shafts. It was largely because of these faster speeds, and increasingly heavy loads the winding engines could transport, that simple tub raising – guided by little more than its attached rope – was replaced by winding cages, these guided by wooden rods fitted into the shaft and later by steel-wire guide ropes.

↑ The original winding engine house and shaft stucture of Bestwood Colliery, Nottinghamshire. *(Steve Bramall/Shutterstock)*

↓ A double-deck cage arrangement, allowing for the lifting of two coal tubs/cars at once. *(Author/Boulton)*

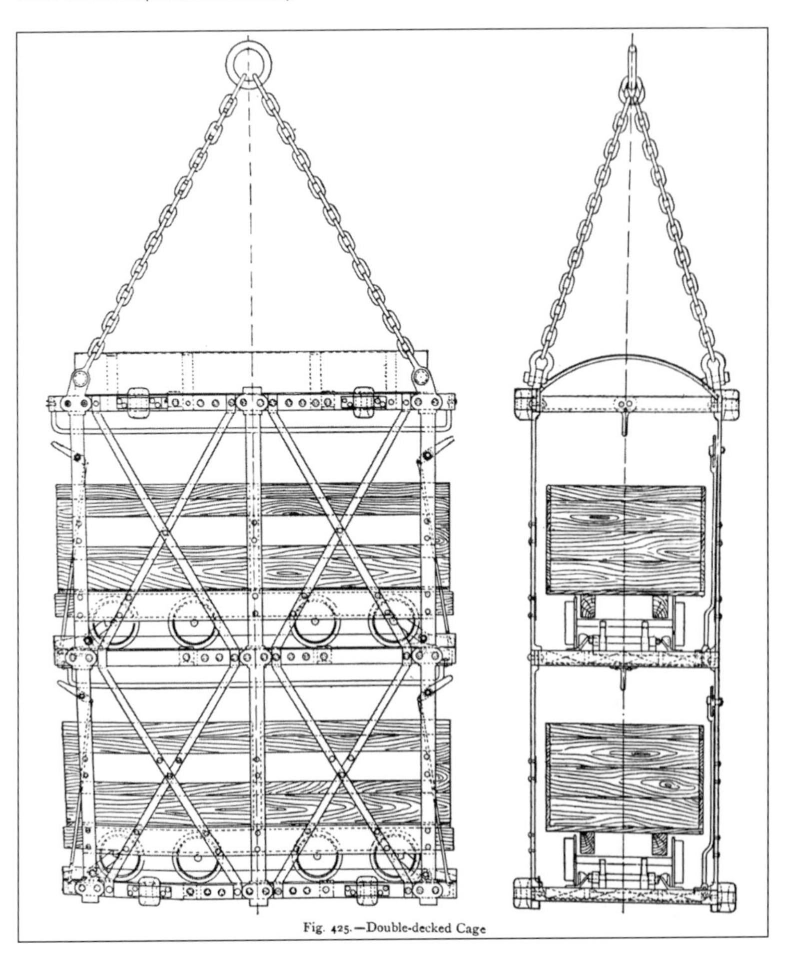

STATE-OF-THE-ART STEAM WINDERS

Steam winding was unusually persistent; it was still the dominant form of winding in the 1950s, and indeed some specimens survived into the 1980s. Mason's 1956 *Deputy's Manual* offers a description of the typical steam winder, in the state of advancement it had reached in the mid-20th century. He states that:

> 'These are not usually geared but coupled direct to the drum-shaft, the drum being on the crank-shaft. The most common type that is the two-cylinder double-acting horizontal engine. The cylinders are up to 42 in. in internal diameter with an 84 in. stroke. They are comparatively simple, reliable and large enough for one cylinder to start the engine against the maximum load. The cranks are set at 90°. Usually the engines are non-condensing, the exhaust steam passing into accumulators for use in turbines. Compound condensing engines economise in steam consumption, but the complication is not in favour of reliability and robustness, and the use of mixed-pressure turbines [. . .] eliminates the necessity for such winding engines.
>
> The cylinders are steam-jacketed and the piston is carried by the crosshead, piston rod and a tail rod to take the weight off the cylinder and reduce wear.
>
> Centrifugal governors act on the inlet valves and shorten the stroke, cutting down the steam supply as the speed rises. At full speed the valves are kept tripped and do not open until the speed falls.'
>
> **(Mason 1956: 428)**

⬇ A fine example of a lattice steel headframe at the Rhondda Heritage Park, Wales. *(Phil Darby/Shutterstock)*

HEADFRAMES/HEADGEAR

Colliery headframes deserve some further consideration before we progress our analysis of winding engines. Often working around the clock, and lifting thousands of tonnes of deadweight every day, headframes are designed to handle three principal stresses:

1. The total weight of the load to be lifted, which includes the coal, cage, tub/car and the winding rope (which when fully extended can weigh more than the other three combined).
2. The weight of the headframe structure itself, and all its associated gear, including pulleys, wheels, guide ropes etc.
3. The angular forces exerted by the pull of the winding engine to raise the load.

To cope with the latter in particular, headframes linked to horizontal winding engines acquired the classic design of four frame legs, two of which stood directly beneath the wheel, straddling the opening of the shaft, and two long bracing back legs extending out to or around the winding house, the whole structure locked together through the use of crossbeams. Certain other design elements gave the headframe greater structural rigidity. The main legs (i.e. those directly under the pulley-shaft), for example, were widely placed at the bottom but converged as close as possible to the outside of the pulleys at the top, directly under the bearings carrying the pulley-shafts (Boulton 1908: 80). This has the effect of shortening the length of the pulley-shaft and thereby increasing strength. It was also imperative that the legs sat on very strong and dependable foundations, ideally large cast-iron shoes attached to blocks of stone, situated on foundations of masonry or concrete if the ground were soft.

During the 19th century, many colliery headframes were constructed of timber, ideally of pitch-pine. During the second half of the century, however, iron and steel headframes were more common as these materials became increasingly available. Not only were iron and steel much stronger than wood, but, properly maintained, they also had far greater durability and resistance to environmental effects. Wood was cheaper, although some experts countered that the cost of maintenance offset any savings.

Metal headgear came in two basic types: 1) those built from solid rolled-steel girders, which were simple and easy to maintain; 2) constructions of latticework, which were lighter and had less wind resistance, but required more demanding maintenance regimes. Importantly, in 1911 an industrial Act prohibited the building of future headgear with timber, and so the metal structures became the standard.

ELECTRICAL WINDING

Electrical winding engines began their steady progressive takeover from steam winding during the first decades of the 20th century. Mason noted in 1956 that: 'Most of the winders are steam driven, and, with their development, braking, governing, control and automatic safety gear have reached a high standard. Electricity, however, is superior to steam as regards efficiency, smoothness of running and adaptability to fully automatic winding. It appears fairly certain that most new winding units will be electrically driven' (Mason 1956: 428). He was right. One of the biggest challenges to the widespread adoption of electrical winding, and indeed electrical power in general, however, was access to sufficient power supply. A select group of wealthy collieries at first built their own personal power stations, but with the government-authorised construction of the National Grid in the 1920s, and its expansion over subsequent decades, the coal-mining industry steadily gained access to a dependable source of external electrical power. Electrical winding also, as Mason indicates, led to the development

⬆ The winding rope runs out from the winding house up to the sheave wheel at Big Pit National Coal Museum, Wales. *(Author/Big Pit NCM)*

⬋ The view from the operator's position in the winding house, looking over at one of the cable drums. *(Author/Big Pit NCM)*

⬇ A perspective view of the Big Pit winding house, the two ropes indicating the double-drum winding mechanism. *(Author/Big Pit NCM)*

of fully automated winding, which from the 1970s and 1980s was also connected to computerised methods of control, giving the operator an unprecedented degree of information, control and awareness of the winding system, with continual feedback on all aspects of its performance.

WINDING MECHANISMS

Winding systems are complex mechanisms, having to negotiate all manner of physical forces, which changed and shifted depending on both the operation the winding system was performing and the stages within that operation. For example, the weight of the load changed significantly during a wind: with the rope full out, the weight exerted at the beginning of a wind could be three times greater than the load that arrived at the pit head, on account of the weight of rope spooled out (mentioned above). This factor required that the system be balanced to compensate for the adjustments.

Balancing could be achieved through one of several mechanical strategies, including:

- Bi-cylindro-conical drum – This system used two winding drums (one for each cage) of variable diameter, winding progressively from the small-diameter portion to the large-diameter portion of the drums, in opposite directions for the two drums, thereby achieving balancing through torque effects.
- Tail rope – The tail rope system attached a rope to the bottom of both cages, in a long loop that dangled loosely down the shaft or ran around pulleys and rollers. As the tail rope weighed the same as the winding

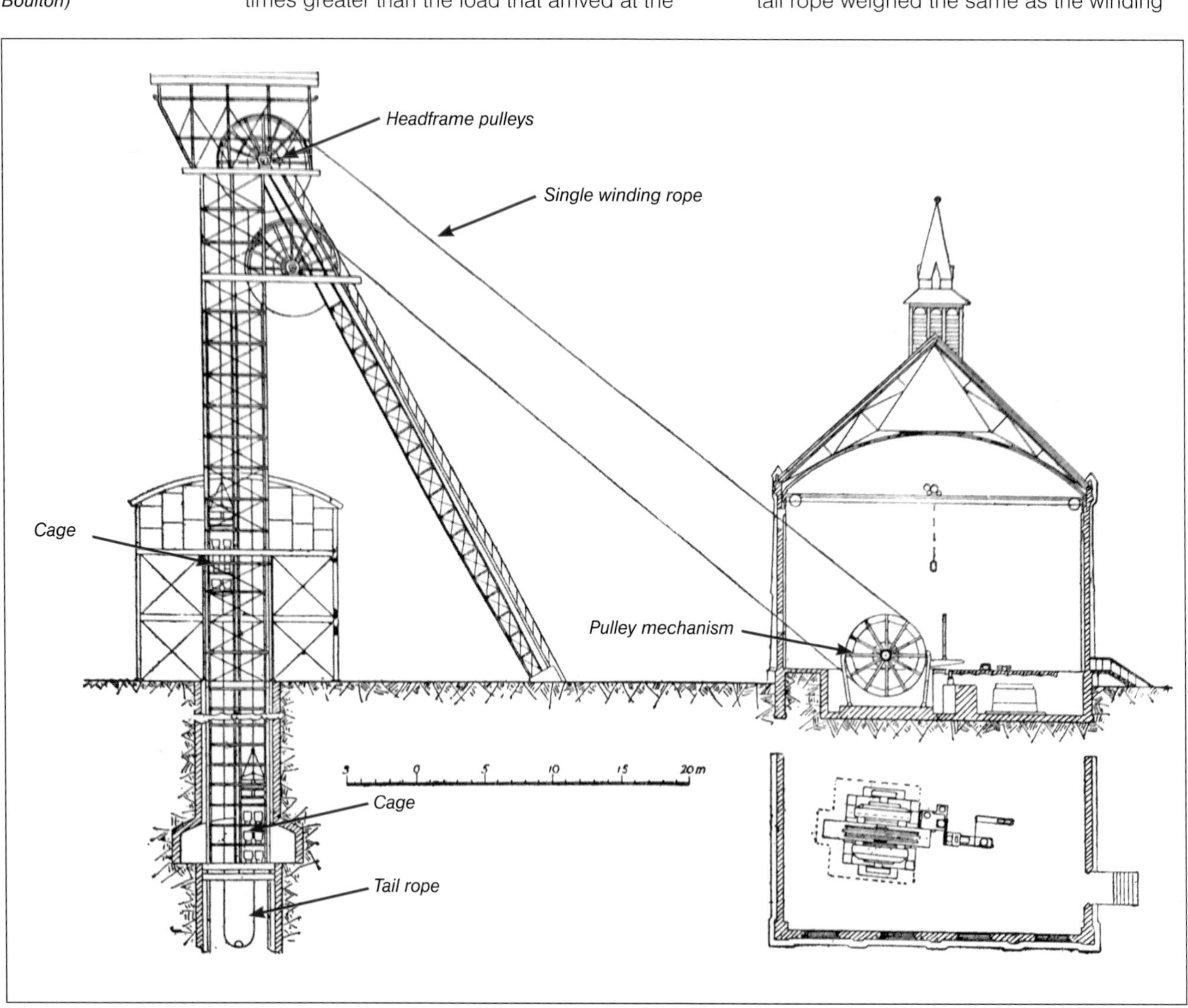

↓ The Koepe counter-balancing system dispenses with a winding drum, substituting it for a winding pulley arrangement. *(Author/Boulton)*

THE JOURNEY IN THE SHAFT

Proper cages for transporting miners and coal came into being primarily in the second half of the 19th century. Some of the earliest systems of taking men to and from the mine workings are positively alarming to modern eyes. The miners might simply descend or rise in the same corves used to transport coal, or stood or hung on slender platforms fixed to a winding rope. One interesting spin on the latter was the 'man-engine', introduced in the 1830s. This consisted of a long and heavy rod that was lifted or lowered by a steam engine on the surface. The rod, which was as long as the shaft was deep, moved up and down rhythmically, and had small platforms fixed to the rod at intervals that were equal to the throw of the engine stroke. Located at the same intervals were fixed platforms in the shaft walls. To use the man-engine, the miner would step on to a rod platform, hopefully with perfect timing, as it rose up to him, then descend and step off on to the next shaft platform down. He would repeat this process until he reached the bottom of the shaft. As the 19th century advanced, there were improvements on this clearly dangerous process; the men might be transferred up and down the shaft in open-sided cages, although these provided little protection from the shaft rushing past. With the faster speeds that came from more advanced steam-powered winding, however, it became imperative to design more functional and protective cages. By the mid-20th century, cages were almost invariably constructed from a steel framework with sides of sheet iron, fitted with gates for entry and access on both sides. Cages could consist of a single deck, but double-, triple- or quadruple-deck cages also came into use, which could carry both men and coal tubs/cars in a single winding.

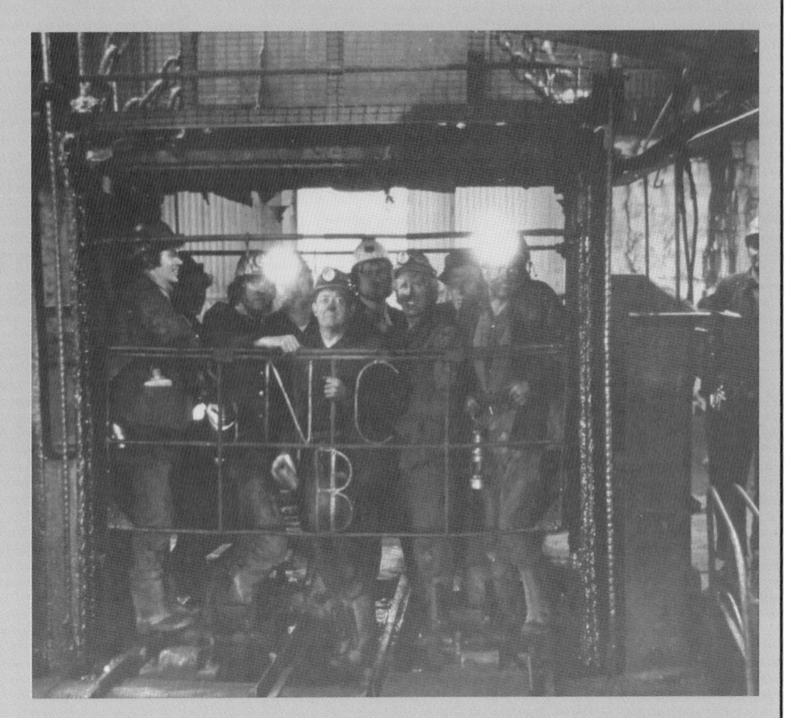

⬆ Miners in a cage ready for their descent to pit bottom. This photograph was taken in what is today the Big Pit National Coal Museum. *(Author/Big Pit NCM)*

⬇ A dual cage arrangement. The levers just visible on the floor of the cages are 'Fisher's catches', used for securing the tubs. *(Author/Boulton)*

🠉 **Set on the surface, this cage has a side-by-side dual track arrangement for transporting two tubs at once.** *(Author/Big Pit NCM)*

🠋 **A shaftsman stands atop the cage to inspect the walls and guides within the pit shaft.** *(Author/Cornwell/Big Pit NCM)*

rope, and moved in sympathy with the winding, the system was balanced and the winder only lifted the net weight of the coal.

Another important system of winding was introduced in 1877 by the German engineer Frederick Koepe. Here, there was only one winding rope, connecting both cages in a continuous loop over the pulley mechanism. The rope did not wind around a drum, and therefore the grip against the winder was entirely a matter of friction, with balancing achieved by the tail rope and counterweights. Koepe winding tended to be used for very deep mines, where other winding systems would require excessively large drums to handle the hundreds of yards of winding cable. The primary advantage of the system was that it needed less power than other winder types to perform its work. The key disadvantage was a reduced rope life, and they can also suffer from rope slip at high loads. In the British coalfields, the first Koepe winder was installed at Plenmeller Colliery near Haltwhistle in Northumberland, in 1914, but it did not achieve wider installation until the late 1920s.

WINDING SAFETY

During winding operations, there are two main safety concerns: a cable snapping, and overwinding, the latter referring to the continual winding of the hoisting rope even after the cage has reached the top or bottom of its travel, usually at excessive speed. Both of these have been negotiated over time with increasingly stringent mechanical, legal and training policies. The former was actually rarely a problem in modern mining, especially with the introduction of high-tensile steel ropes with breaking strengths well in excess of the loads carried.

This being said, the winding wires would, over their lifespan, start stretching from the day that they were installed and began operational use. The changes in length that resulted could be compensated for, to a degree, by adjusting a distributing plate or 'kidney plate', which provided fine adjustments to ensure that the cage landed as level as possible with the pit bank. It was a metal plate that formed the link between the chains attached to the roof of the cage (one at each corner, plus two safety chains) and the metal blocks that connected to the detaching hook (see below), which in turn linked to the capping. (The capping was the point at which the end of the rope, its strands flared out and set in a cone of hardened resin or metal – the cap – connected with the suspension gear on top of the cage.) The kidney plate was drilled through with three linkage holes, which could be used in sequence to adjust for the wire stretching 3in (76mm), 6in (152mm) and 9in (229mm); beyond these limits, the entire rope would need to be replaced, which was done at regular intervals anyway depending on the use and lift-loads imposed.

OVERWINDING

Although snapped cables were rare, overwinding was a problem that caused intermittent but often serious loss of life and limb (my own grandfather attended the aftermath of a serious cage overwind accident in the 1930s). Overwindings were violent accidents, resulting in either a cage of people smashing into the headgear or crashing into the bottom of the shaft at velocity, or both. Mining history is littered with such accidents, one of the most prominent in recent history occurring at Markham Colliery at Staveley near Chesterfield, Derbyshire, on 30 July 1973. On this occasion, a fractured brake rod caused an overwind that killed 18 miners and seriously injured a further 11. Following this accident, it became a legal requirement for all collieries to fit their winding gear with the Ormerod detaching hook (described on page 89).

Overwinding accidents had three main causes: 1) human error; 2) brake failure; 3) the winding engine running out of control. Human error could largely be limited by proper training, especially in speed control and braking. In a traditional winding station, with mostly analogue displays and controls, the operator would perform a winding while looking at circular metal dials, on which markings were displayed that denoted the shaft position of the cages under his control; moving arrows indicated the position of the cages during a winding relative to these markings. Specific zones on the dial designated permissible speed limits; thus in the middle of the cage's journey it might be doing

↑ The haulage system inside a mine would run tubs and cars directly into position for cage loading at pit bottom. *(Author/Cornwell/Big Pit NCM)*

↓ A diagram showing a system of cage rope capping, with the collars A, B and C helping to prevent slippage. *(Author/Boulton)*

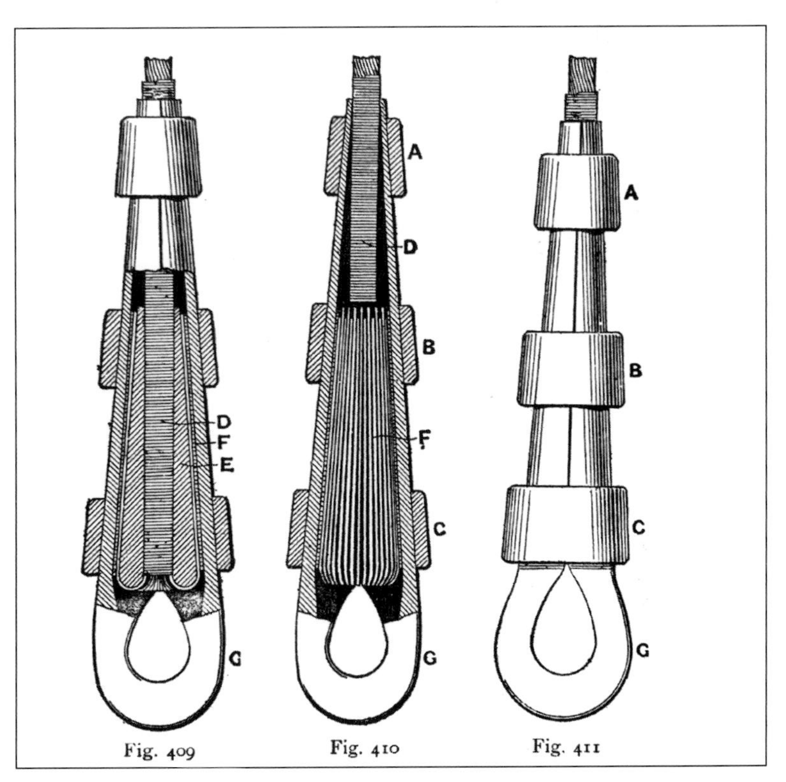

20ft/sec (6m/sec). The operator would steadily decelerate the speed as the dial indicated the cage's approach to the shaft top or bottom, down to about 3ft/sec (0.9m/sec) on the approach to stopping. Braking was performed by a manual brake on the drum or by dynamic resistance braking imparted by the engine.

This procedure could be faithfully performed across hundreds of pits and tens of thousands of lifts every year. Yet human error and mechanical failure could never be entirely eradicated, thus during the late 19th and 20th centuries a steadily expanding sequence of safety mechanisms were introduced to provide multiple preventive back-ups against disaster.

Specifically, these were:

- **Slack rope** – In a classic slack-rope device, a cable connected to an automatic engine braking system was strung across the aperture through which the winding rope entered the winding house, the winding rope sitting directly beneath the taut rope. In an overwind situation, the winding rope would often go slack as the drums ran faster than the rope could spool; this caused the rope to drop against the cut-off cable, bringing the overwind to a halt. Note, however, that a slack rope device does not have to be a physical cable; it can be any device that is fail-safe

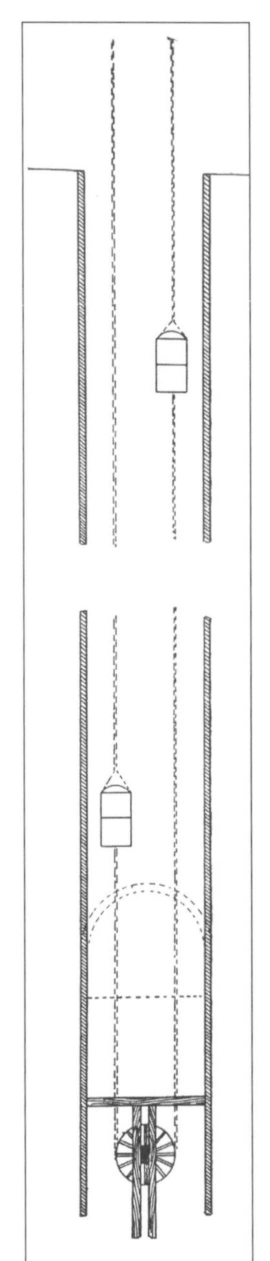

⬅ **A tail-rope counterbalance system, with tail ropes hanging from the underside of each cage via capping.** *(Author/Boulton)*

⬇ **A close-up of a Lilly duplex controller safety system, showing one of the cam dials on the side.** *(Author/Big Pit NCM)*

⬇ **The dial at top indicates the position of the cage in the shaft, with the coloured nodes indicating speed and braking states.** *(Author/Big Pit NCM)*

⬇⬇ **Electrical winding took over from steam winding during the first half of the 20th century.** *(Author/Big Pit NCM)*

and can detect a slack rope situation, and several designs have been used.

- **Lilly duplex controller** – This mechanism used two cam dials, one for each direction of cage motion, mounted on hubs and driven by the movement of the winding drum. On the top of the mechanism were two centrifugal governors, driven by a shaft from the winding drum, so that the speed at which the governors span reflected the speed of the cages. If the speed exceeded pre-set limits (which varied according to whether people or coal was being hauled), the information was mechanically transmitted from the governors to the cams, which triggered an alert as a first response, then a power shut-down and emergency brakes.
- **Magnetic cut-offs** – Magnetic sensors were located on the headframe above the cage's normal limits of travel; moving past these would trigger the emergency brakes.
- **Ultimate switch** – Further up the headframe, a projecting lever – which would be struck and pushed up by the cage canopy in the case of an overwind – would also trigger the automatic safety mechanisms. A rack system running up the interior sides of the headframe provided a surface against which the cage would lock itself, to prevent it from dropping back down the shaft.
- **Detaching hook** – The ingenious Ormerod detaching hook was patented by 1867 by Edward Ormerod, an engineer at Gibfield Colliery, Atherton, Lancashire, and by 1954, 10,000 had been installed. Indeed, after the Markham disaster of 1973, it became a legal requirement for all underground mines to fit these to their winding gear. In this system, the rope of the winding gear passed through a bell-mouthed cylinder. Located just above the cage chains, the detaching hook consisted of three movable

A diagram of the Omerod detaching hook, one of the most important safety innovations in mining history. *(Author/Boulton)*

An illustration of the Omerod detaching hook with the rope detached (left) and the hook ready for lowering (right). *(Author/Boulton)*

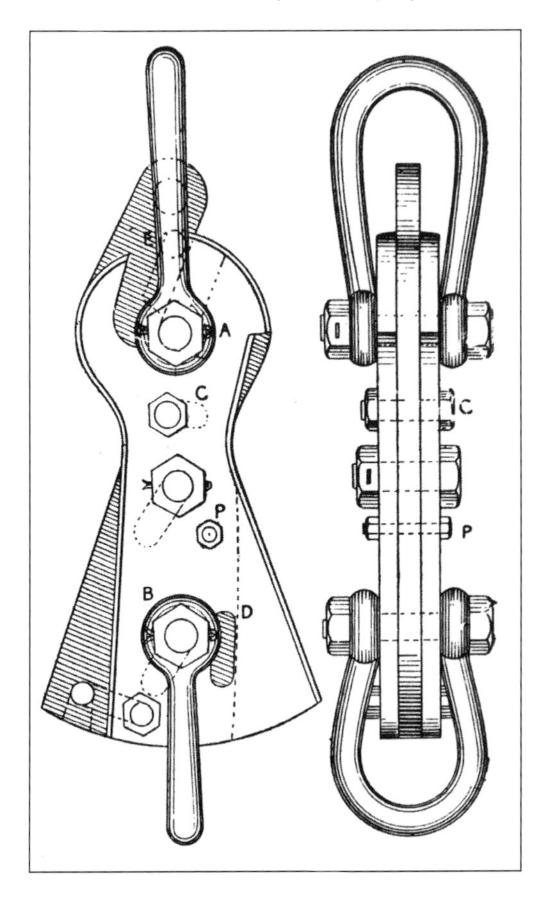

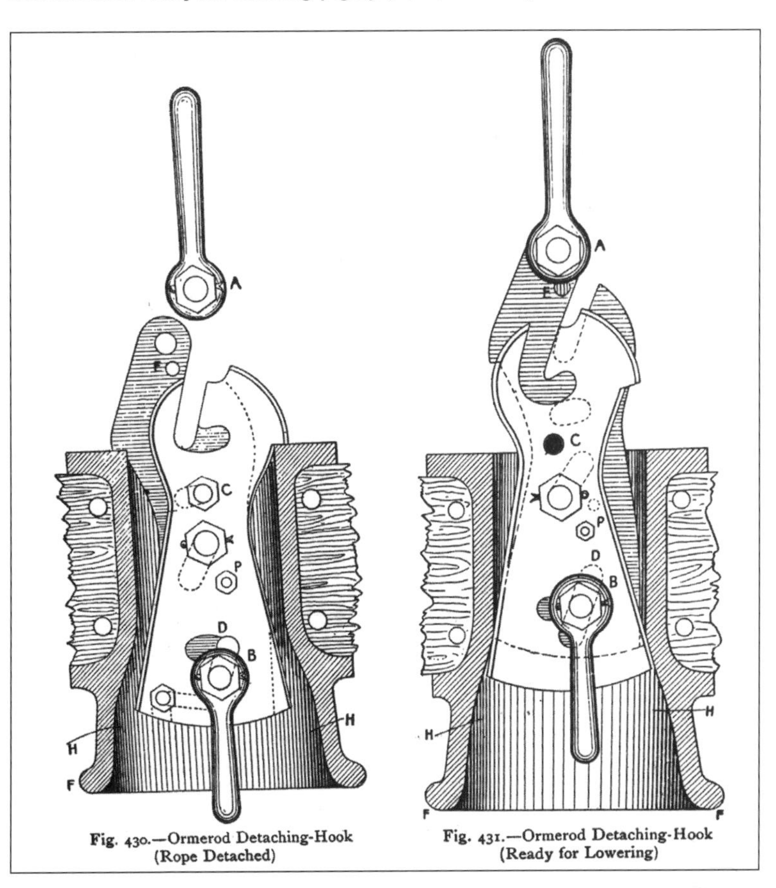

Fig. 430.—Ormerod Detaching-Hook (Rope Detached)

Fig. 431.—Ormerod Detaching-Hook (Ready for Lowering)

➡ **In this view above a cage, we can clearly see the detaching hook shearing plate, projecting out about a third of the way down from the top of the link.** *(Author/Big Pit NCM)*

⬇ **A cross-section of surface-level hydraulic cage loading and unloading apparatus, with the cage rising above ground in the centre and hydraulic rams pushing them out.** *(Author/Boulton)*

plates held on by a central bolt and secured in the closed position by a small metal pin. In the case of an overwind, the hook would be dragged up through the bell plate, an act that forced the wings of the detaching hook inwards, shearing the metal pin and releasing the rope above, preventing the cage smashing into the top of the headgear; two projections automatically opened out on the top of the detaching hook, these now resting on top of the bell and preventing the cage from falling back down the shaft.

- **Computerised safeties** – From the 1980s, computerised safety systems were introduced into winding mechanisms. These analysed multiple aspects of winding safety and performance in minute detail, and could be programmed to trigger braking and power shut-downs when necessary.

With the advent of this bank of safeties, shaft transit became one of the least risky aspects of a miner's working day.

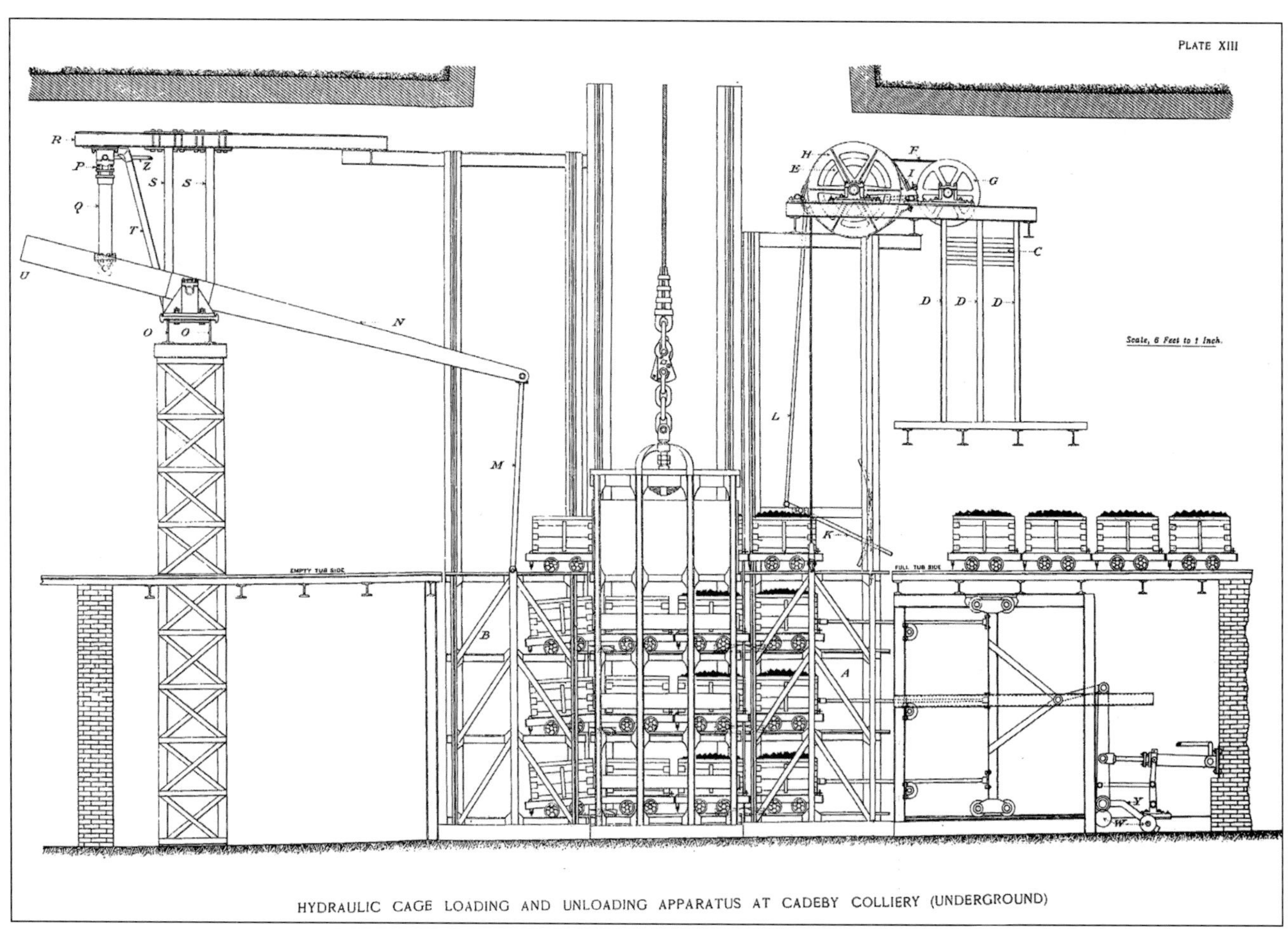

WINDING SIGNALLING CODE

The three people in charge of winding operations are:

1 the banksman (the person in charge of supervising cage operations at the top of the shaft)
2 the onsetter (his counterpart at the bottom of the shaft)
3 the winder operator

In the days before electrical installations, the way they communicated was via a pull-rope system – a tug or specified series of tugs on the rope at the top were physically received at the other end of the rope at the bottom – or sometimes, with shallow shafts, by striking a metal plate with a hammer. With electricity came other means of communication, such as telephone, buzzers and a numbered light board with buzzer buttons, known colloquially as the 'knocker'. The following is the signalling code used on this machine, the number indicating the corresponding button to press.

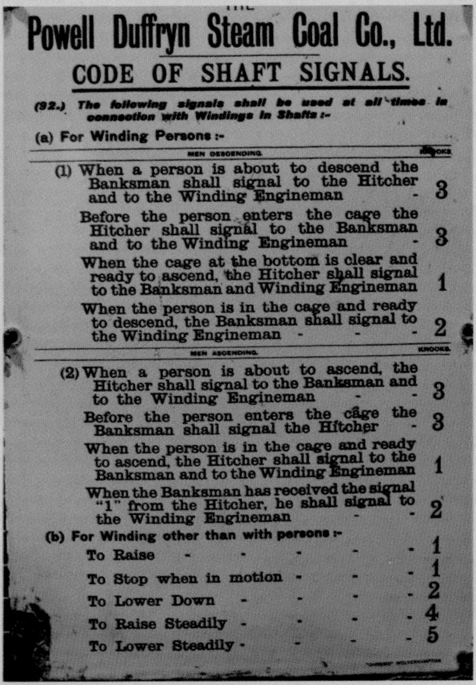

THE Powell Duffryn Steam Coal Co., Ltd.

CODE OF SHAFT SIGNALS.

(92.) The following signals shall be used at all times in connection with Windings in Shafts :-

(a) For Winding Persons :-

MEN DESCENDING.	KNOCKS.
(1) When a person is about to descend the Banksman shall signal to the Hitcher and to the Winding Engineman	3
Before the person enters the cage the Hitcher shall signal to the Banksman and to the Winding Engineman	3
When the cage at the bottom is clear and ready to ascend, the Hitcher shall signal to the Banksman and Winding Engineman	1
When the person is in the cage and ready to descend, the Banksman shall signal to the Winding Engineman	2

MEN ASCENDING.	KNOCKS.
(2) When a person is about to ascend, the Hitcher shall signal to the Banksman and to the Winding Engineman	3
Before the person enters the cage the Banksman shall signal the Hitcher	3
When the person is in the cage and ready to ascend, the Hitcher shall signal to the Banksman and to the Winding Engineman	1
When the Banksman has received the signal "1" from the Hitcher, he shall signal to the Winding Engineman	2

(b) For Winding other than with persons :-

To Raise	1
To Stop when in motion	1
To Lower Down	2
To Raise Steadily	4
To Lower Steadily	5

WINDING PERSONS

Descending

Stage	Signal number
People are about to descend. The banksman signals the onsetter and the winder.	3
Before any people enter the cage, the onsetter signals the banksman and the winder.	3
When the cage at the bottom of the shaft is clear and ready to ascend, the onsetter signals the banksman and the winder.	1
When people enter the cage and are ready to descend, the banksman signals the winder.	2

Ascending

Stage	Signal number
When people are about to ascend, the onsetter signals the banksman and the winder.	3
Before any people enter the cage, the banksman signals the onsetter.	3
When people are in the cage and ready to ascend, the onsetter signals the banksman and the winder.	1
When the banksman receives the signal '1' from the onsetter, he signals to the winder.	2

WINDING COAL OR MATERIALS

Stage	Signal number
Raise up	1
Stop	1
Lower	2
Raise steadily	4
Lower steadily	5

(Source: Mason 1956: 435)

⇦ A metal plate displays the code of shaft signals for use between the banksman, onsetter and winder. *(Author/Big Pit NCM)*

Mechanisation: COAL EXTRACTION AND PROCESSING

It was during the 19th century that engineers began to make their first attempts at mechanising coal-cutting, recognising that the processes of hand-got coal were limited by the dynamics of the human body. The first coal-getting machine was seen in the 1760s – it was essentially a solid iron version of a miner's pick mounted to a frame and racked back and forth by two men operating a geared winding wheel. Through the basic application of physics, the machine notionally did have more efficacy than hand-wielded tools, but in reality it was awkward to use, and had few advantages over regular picks and hammers. What was needed was to supply such tools with power.

⬅ **A fitter cleans and sharpens the picks of a roadheader.** *(Author/Cornwell/Big Pit NCM)*

⬆ **The 'Iron Man', an early type of hydraulically powered coal-cutting machine, introduced in the 18th century. This example dates to 1864.** *(Author/Big Pit NCM)*

⬆ **A front view of the pick of the 'Iron Man', which was swung in a horizontal plane to undercut the coal seam, albeit not terribly efficiently.** *(Author/Big Pit NCM)*

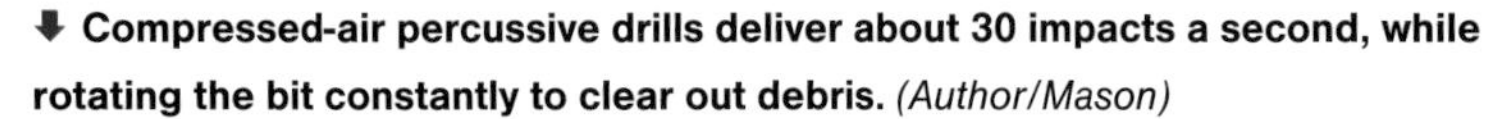

⬇ **Compressed-air percussive drills deliver about 30 impacts a second, while rotating the bit constantly to clear out debris.** *(Author/Mason)*

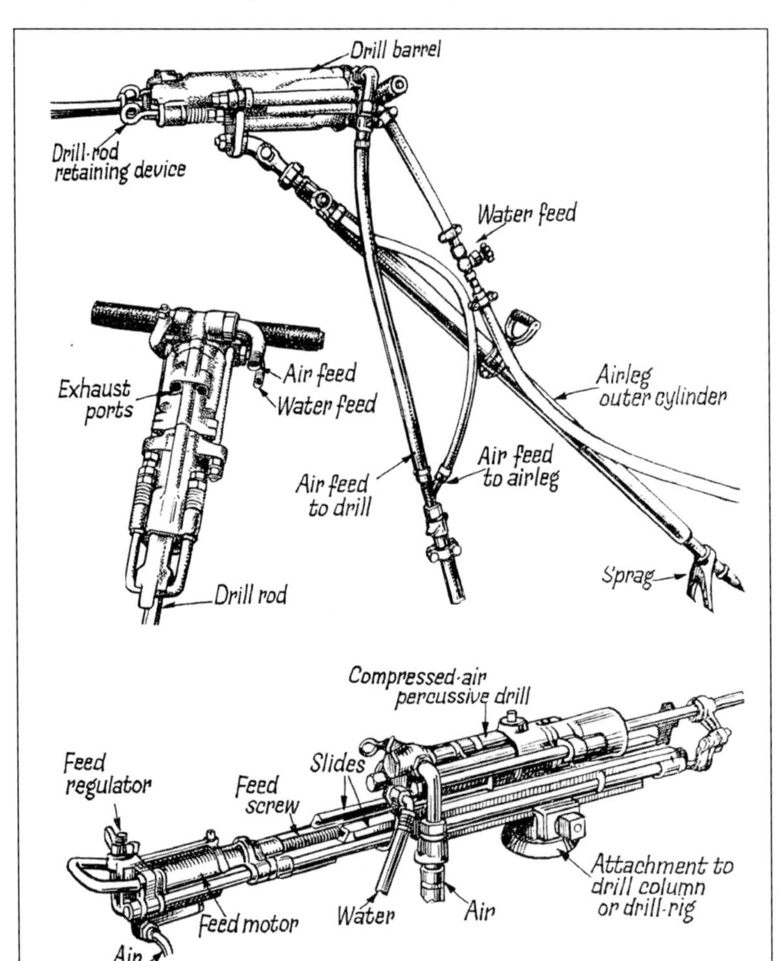

The true step forwards in mechanised coal-getting came in the later part of the 19th century, when compressed air came to the coalface, followed later in the century by electricity. Using these two power sources, new generations of coal-cutting and heading machines came into the hands of miners, progressively stepping up the productivity of the individual and the team at the coalface. In this chapter, we will review some of the key categories of machines, from simple pneumatic picks through to the modern generations of shearer-loaders.

POWERED PICKS

Mechanised picks and hammer drills, powered by compressed air, were used in British coal mining from the late 1800s until the present day. These were either hand-held tools or, for improved results on the coalface, could be mounted on frames to apply heavier and more consistent pressure. (The earlier types of pneumatic picks were weighty objects in themselves, so often needed support anyway.)

The pneumatic picks were used either for undercutting or for bringing down an undercut seam, in places where other types of coal-

← Two miners in the early 20th century use an Ingersol 'Radial' Machine to cut a heading, the machine steadied by a vertical fitting between floor and roof. *(Author/Boulton)*

↓ A Siskol rotary coal cutter. The head would be swung in an advancing arc to undercut the coal or rock face. *(Author/Big Pit NCM)*

getting machinery couldn't be brought into play. There were also other economic benefits of extracting coal by this method, such as reductions in shotfiring and cleaner and larger pieces of coal with less wastage, which could also reduce the coal processing costs on the surface. As these tools did not have their own integral power supply, however, the coalface had to be rigged with regular compressed-air feed points to supply multiple picks. A 2in (50mm) diameter compressed-air main was laid along the face near the conveyor. This featured regular flexible joints, such as Carlton or Unicone types, and was provided in lengths of 6–9ft (1.8–2.7m), so it could be dismantled quickly. (If the pipework was to supply more powerful coal-cutting machines, then they were a minimum 76mm (3in) in diameter.) At regular intervals along the face, the pipe had a feed point to supply the tools, including a two-way feed to serve two picks at the same time. The valves opened automatically when the air-pick hose was screwed into position and closed when the hose was unscrewed.

Although the picks were certainly effective, they could be hard on the user in terms of noise, dust and vibration injuries transmitted to either the hands or the knees (knees were often

The body of an electic coal drill, a dramatic improvement over the days of the manual pick. *(Author/Big Pit NCM)*

used to provide extra pressure behind the pick cut). Special pads were developed to protect hands and knees, and in time the picks came to have better internal anti-vibration mechanisms, and thus were kinder on their users.

CUTTING MACHINES

While much work could be done on a short stretch of coalface by hand-held tools, a lengthy longwall face or extensive pillar-and-stall workings required more industrious mechanisation. Prior to the advent of shearers and ploughs that cut the coal directly perpendicular to the coalface, the most important coal-getting machines were those compressed-air or electrically powered devices that mechanised the process of undercutting. These were divided into three principal categories, all of which have remained in use throughout the modern era of mining:

1 disc/wheel cutters
2 bar cutters
3 endless-chain

DISC/WHEEL CUTTERS

Disc/wheel cutters were, by the early years of the 20th century, the most prolific of the cutter types. In simple outline, the machines consisted of a frame set on wheels or skids, the frame containing air cylinders or the electric motor (depending on the power supply type), gearing wheels and a rope drum used for dragging the machine forwards across the length of the face. A powered disc cutter – a flat disc of metal with cutting teeth running around its circumference – projected from one side of the frame (the cutting side).

Depending on the power and size of the machine, a disc cutter could make a cut

The Jeffrey longwall machine had a horizontal plane cutting disk, with slight oscilation via a rocker-arm. *(Author/Boulton)*

of up to 7ft (2.1m) in depth, the cut height being 3–5in (76–127mm). Because of their design, disc cutters had a high degree of cutting efficiency. *Practical Coal-Mining* (1908) explained that in the compressed-air machines, the disc made about one revolution to about 20 revolutions of the engine shaft, with a higher ratio – about 65–70 revolutions of the engine shaft to one of the disc – in electrically powered machines. The discs would typically revolve at speeds of anywhere between 10 and 30 revolutions per minute (the larger the machine, the slower the revolutions), although there was a machine made by Rigg and Meiklejohn that delivered 60rpm. Such speeds dramatically accelerated the undercutting process.

To draw the disc cutter along the coalface, a rope was tied to the front of the machine and passed around a small sheave or snatch block, which was in turn chained to a screw jack, a length of rail wedged between the floor and the roof, or another similarly solid restraint. The rope went around the sheave and into the drum winding mechanism on the machine, which pulled the machine forwards. The wheels of the disc cutter generally ran on rails; *Practical Coal-Mining* specifies that just three or four lengths of rail were typically used, those behind the machine being lifted up and laid in front of the machine as it advanced. The outer rail had to be kept rigid by the use of props, to prevent the natural outward thrust generated by the machine's cutting action from pushing it away from the face.

BAR CUTTERS

The bar cutter had, instead of a cutting disc, a rotary powered bar, studded with the cutting picks. In more detailed description, the machine had three main units: 1) the driving unit (which contained the motor); 2) the haulage unit to pull the machine across the face (this part also included the operator's controls); 3) the gearhead, which featured a rotary cutting bar. The chief advantage of the bar cutter was its versatility and ease of use compared to the disc cutter. It could be employed in smaller spaces, and because the bar was hinged it was better at cutting out on corners. It took up less space than the disc cutter, and, furthermore, the miners operating the cutter could insert sprags into the cut just behind the face of the cutter's advance. On the downside, bar cutters could be more susceptible to

A Gillet & Copley disc cutter, patented in 1868 but still in use in mine work in the first half of the 20th century. *(Author/Boulton)*

⬆ The Hurd Bar Machine was a bar-cutter type, the cutter arm both rotating and moving in and out to avoid clogging. *(Author/Boulton)*

➡ A close-up of a powered rotary drill head. *(Author/Big Pit NCM)*

➡ A BJD 'Ace' continuous-chain coal-cutter, commonly used in mining from the 1940s to the 1960s. *(Author/Big Pit NCM)*

damage because of the pressure placed on the rotary bar, and they also removed less dirt from the cut compared to the disc machines. The bar cutter was drawn across the seam via a rope, attached to a distant anchor point, winding the machine across via the haulage unit.

ENDLESS-CHAIN CUTTERS

The most popular types of coal-cutting machine became the endless-chain variety. At a glance, these cutters in many ways appear as a large chainsaw, the cutting jib, in the most simple forms, projecting horizontally and in a flat plane from the side of the machine, with an endless pick-studded chain running around the outside of the jib.

Different types of jibs (and picks) could be fitted on to the machines to suit different cutting challenges. The majority of cutting with these devices was done with straight jibs, although curved jibs were also used from the late 1950s. These jibs were shaped to make both horizontal and, deeper inside the seam, vertical cuts, the extended cuts either going up into the seam from the bottom or down into the seam from the top. Making cuts in this way reduced the amount of shotfiring required, or could even obviate it altogether. Chain cutters could also be easily configured to make 'middlecuts' (cuts in the middle of the seam) and 'overcuts' (cuts at the top of the seam), if the particular conditions of the seam warranted it. Another advantage of the endless-chain cutters was that their action was effective at cleaning out dust and other detritus from the cut as the machine was drawn across the face.

HANDLING CUTTING MACHINES

The following advice comes from a 1950s NCB manual, *Operation of a Longwall Coal Cutting Machine*, in the section 'Hints to Cutter Men':

'The difficulties which arise when cutting are of several kinds, the following are the most common.

1. *The anchor prop may tend to pull out. This is most usual in seams with soft floors. Have a "shoe" arrangement made to fit to the foot of the anchor prop so that the prop does not dig into the floor. Ensure that the anchor prop is set directly in line of pull; if it is set "askew" the foot of the prop will tend to ride free.*
2. *The cutting end may tend to leave the face. When this happens, examine the picks, since these will probably have become blunt. Change the picks and set stays or sprags against the haulage end so that as the machine passes the sprag it is kept close to the face.*
3. *When the floor is uneven pack the hollows with wood or steel bars, so that the machine will ride over them, to keep the cut in its proper position.*
4. *The machine may stall.*
   A good cutterman can tell when a machine is becoming overloaded by the change in note of the motor. The variation in note gives warning to reduce or stop the feed, to change the picks or to clear away the gummings from the gearhead.
5. *The jib may tend to ride above or below the correct position of the cut. This may in some cases be corrected by running the machine over a small piece of wood. If this does not cause the machine to cut correctly it indicates that the chain may require tightening or that the picks are not set properly. If the machine tends to rise the top picks are set too far out; if the machine tends to dig, the bottom picks are set too far out.*
6. *Excessive cutting speed may cause the jib to move into the softer parts of the coal or dirt being cut and therefore to rise or dip.*
7. *The gummer may become jammed. If so, stop the machine and look for and remove the cause. Momentarily reversing the motor may remove the obstruction.*
8. *The chain may break. If this happens, stop the machine, remove the gummer and free the jib. If necessary draw the jib out of the cut. Repair the broken chain.*
9. *The jib may become fast. If it does, reversing the chain may free it. If this does not help, pick away the coal to free the jib.*

Shots should never be fired directly over a jib since permanent damage to the machine is likely to be caused.

10. *With compressed air machines the use of wet air, overloading or throttling of the air may cause ice to be formed in the rotors in such quantities as to lock them. This may be prevented by the use of water traps in the pipe system or by keeping the rotors revolving slowly whilst other work is being carried out. Never switch full air pressure on to a machine running light.*
11. *The haulage rope may fall and be caught by the cutting picks when slewing the jib. The rope should always be kept taut.*
12. *The trailing cables connected to the machines may foul the cutter chain or become strained at the point where they are attached to the socket. Make sure they are so placed that this cannot happen.'*

(NCB 1953c: n.p.)

⬆ A miner uses a control unit to direct the operation of a coal shearer-loader machine. *(Author/Cornwell/Big Pit NCM)*

⬇ The Meco-Moore cutter loader, featuring two parallel horizontal cutting jibs and a vertical shearing jib. *(Author/Big Pit NCM)*

Two miners, working a very low seam, operate a gathering-arm loader to transfer the cut coal back for loading. *(Author/Big Pit NCM)*

CUTTER-LOADERS

From the 1930s, the coal mining industry was increasingly eager to develop systems of mechanisation that enabled 'continuous' mining, i.e. mining that was free from the cyclical shift pattern of cutting, filling and face preparation. What we see, therefore, is a steadily more intimate marriage between cutters, loaders, conveyors and props, eventually blending into one interlinked process that could perform all these actions simultaneously. In the following chapter, we will look more closely at roof supports, including automated systems, while here we will focus specifically on cutter-loader development.

MECHANICAL LOADING

'Loading devices' refer to machines that mechanically load the loose coal and transfer it to the haulage system, automating a process that generations of miners had performed by hand and shovel. These machines could work in tandem with the cutter, loading as fast as the coal was brought down. Before looking at cutter-loaders, we therefore need to clarify the mechanical loading systems available in their developed state by the 1950s and 1960s. These were:

Flight loaders – A flight loader utilised metal 'flights' (metal scraper bars) that projected out from the side of a jib and moved in a rotary motion, scraping up the coal cut from the face and pushing it forwards on to ramps that led up to a conveyor. The Huwood loader was one of the most popular types of flight loader, with a front-mounted system of flights that moved horizontally. Driven by either compressed air or electricity, this machine could work over the same gradients as the coal-cutter, and was pulled across the face by two haulage wires attached to winches on the rear face of the machine. To achieve smooth movement and stay close to the face, one of these wires had to be installed at the rear of the cut before the coal was brought down. If the rope were severed during this process, there was a mechanical means of pushing rather than hauling the machine.

Gathering-arm loaders – These machines were mounted on crawlers, which pushed the machine forwards and drove a long ramp, leading up to the conveyor, into the cut coal. As the coal ascended the ramp, two powered arms, moving constantly via a revolving disc mechanism, literally pushed and guided the coal mechanically up the conveyor. The

⬅⬅ **The scraper chain of a Joy-Sullivan gathering-arm loader. Gathering-arm loaders were generally used for headings and roadway-making, but also on longwall faces.** *(Author/Big Pit NCM)*

⬅ **The gathering arms of a loader, which swept forward and back alternately to push the coal onto the conveyor.** *(Author/Big Pit NCM)*

Deputy's Manual notes that this type of machine category had a high rate of loading, providing the faster speeds in flat to moderate gradients (Mason 1956: 505).

Duckbill loader – The principle of the vibrating duckbill loader has already been outlined, the machines using reciprocating motion to shake the coal in the desired direction. Duckbill loaders worked best travelling up an incline – they were not suitable for downward-facing work – and they moved via a sliding telescope framework. Angled troughs could be fitted that allowed the shaker conveyor to operate around angles of up to 90 degrees.

Mechanical shovels – The mechanical shovel was rather like a vehicular digger. It ran either on rails or crawler tracks, scooping up coal in its bucket and then swinging it over the top to discharge it into an adjacent tub.

MECO-MOORE CUTTER-LOADER

In the 1930s there arrived the first practical machine that performed the duties of cutting and loading within the same unit. This was the Meco-Moore, invented by engineer M. Moore, and manufactured by the Mining Engineering Company (MECO) of Worcester. The machine was powered by two motors, one driving a horizontal endless-chain cutting jib and the other the loading mechanism. The latter consisted of a loading bar at floor level, which lifted the coal through a spiral motion and transferred it on to a loader conveyor, which sat at a right angle to the coalface, and which transferred its load in turn to the face conveyor.

The Meco-Moore at first had a two-directional working process. On its first pass

⬇ **A miner guides a cutter-loader forward, with the horizontal mid-height cutting jib ready to bite into the seam.** *(Author/Big Pit NCM)*

➡ **A coal plough shearer blade section, attached directly to the powered loader system.** *(Author/Big Pit NCM)*

down the coal seam, the horizontal jib would insert cuts into the coal. When the machine had reached the other end, the coal would be blasted, after which the machine would make the return journey; during this transit, the loading system scooped up the coal for haulage. In the 1940s, however, the machine was much improved by the firm of Anderson Boyes Ltd, who created the AB Meco-Moore, which featured the addition of another horizontal cutting jib that cut at the mid-point, as well as a large vertical shearer jib next to the loader assembly. This meant that the machine could undercut, shear and load at the same time in the same pass.

⬇ **Anderson Boyes machines were amongst the most popular brands of coal-cutters of the 20th century.** *(Author/Big Pit NCM)*

Thus the era of the cutter-loader had arrived. However, the Meco-Moore was far from perfect. The depth of the cut it produced had a tendency to weaken the roof of the coal seam, and it also produced an excessive amount of dust (a problem that was a common accompaniment to mechanisation). Meco-Moore use peaked in 1957, with 165 of the machines in operation, but by that time several other machines had entered the arena as competitors, with evocative and energetic names such as the Logan Slab-Cutter, the Uskside Miner and the Gloster Getter, working on similar principles.

COAL PLOUGHS

There were other innovations, such as coal ploughs. These were track-clearing ploughs fitted with vertical cutting blades. The plough was drawn across the face by a haulage engine, cutting a slice of coal as it went, with the plough blades directing the coal on to the conveyor system. These machines could cut and load when moving in either direction and could work in gradients of 1:2. One of the most distinctive of the plough types was the Samson Stripper, developed from the early years of World War II by the firm of Mavor and Coulson. This was a self-propelled plough cutter-loader, which pushed itself along the coalface via an integral travelling jack propelled by a piston and rod. Fitted with four vertical cutting blades on each end, it could cut the full height of the seam, while also pushing the coal on to the loader. It was, however, beset with technical difficulties and didn't take off in the significant numbers its manufacturer hoped for.

ANDERTON SHEARER-LOADER

In the war years of 1939–45, a time of insatiable demand for coal, investment in mechanisation was heavy within the British coal industry. It was during this era that the AFC was developed, entering service in the UK shortly after the resolution of the conflict. The advantage of the Panzer conveyor was not only its haulage capacity, but also its flexibility, which meant that even as a cutter progressed

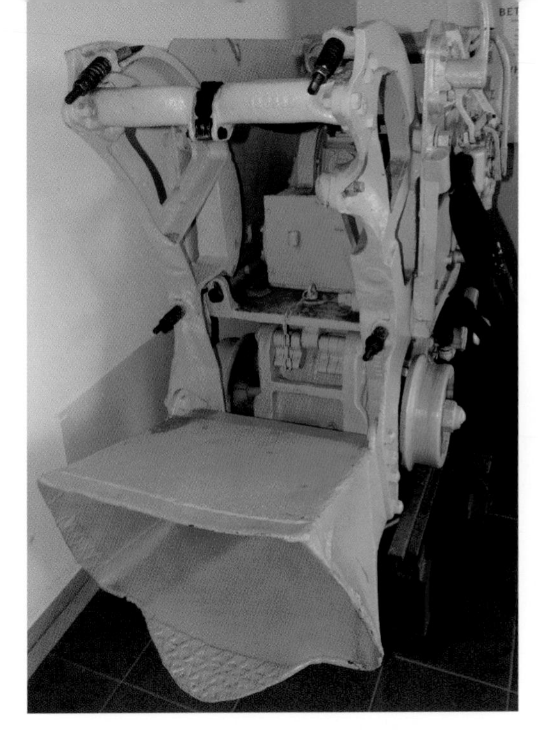

ahead the conveyer could be pushed up to the seam in readiness for the return cut. Then, in 1946, the first hydraulic props were introduced, and two decades later came fully powered variants of such props, revolutionising coalface support (see Chapter 5).

One of the signal events in the mechanisation of coal-getting, however, was the invention of the Anderton shearer-loader. This was developed by NCB official and engineer James Anderton, working in the Lancashire coalfields. His innovation was to reinvent the disc cutter principle, but instead of mounting the disc horizontally, he stood it vertically, so that when rotated it cut along the face of the seam. Also, instead of a single flat cutting disc, Anderton fitted a rotary drum with multiple bands of cutting picks looping around its outer surface, giving it an operational performance that has been likened to that of a bacon slicer. Behind the shearer drum were plough vanes that directed the cut coal on to the AFC. Furthermore, the shearer itself was physically connected to the movement of the AFC – the conveyor therefore provided the cutter's haulage – thus binding the cutting,

↖ An Eimco 'Rocker-Shovel', used to load coal into tubs from the 1940s–60s. *(Author/Big Pit NCM)*

↑ A state-of-the-art Eickhoff coal shearer, set on its oscillating arm. A Russian Eickhoff SL 900 shearer-loader produced 1.57Mt of raw coal in a year, mined from a single face. *(Author/Big Pit NCM)*

← A low-down view of the 'Panzer' conveyor of a shearer-loader system, looking up at the hydraulic supports. *(Author/Big Pit NCM)*

THE MODERN MINING PROCESS

An excellent view of modern coal cutting, with self-advancing supports holding up the roof and a shearer-loader system simultaneously cutting and collecting the coal. *(Author/Cornwell/Big Pit NCM)*

With the advent of the shearer-cutters, the AFC and the powered roof support, coalface operations reached a state of fluid efficiency. Summarising the steps of production, they were broken down as follows:

1 The gates either side of the coal seam were advanced, cut either by roadheaders or by the extended sweep of the shearer-cutter arm. The coal seam was now exposed for working.

2 The cutter-shearer and the AFC were brought up to the face, and made a pass along the coal seam, cutting and loading the coal in a simultaneous action.

3 Typically after two passes along the coalface with the cutter-shearer, the hydraulic props were advanced forwards to support the exposed roof of the coal seam. The sides of the gates were supported with packs (the following chapter explains packs in detail). The goaf behind the hydraulic props was allowed to collapse.

4 The gates were now advanced and the process could begin again.

loading and haulage functions into one seamless whole.

The Anderton shearer-loader's efficiencies were immediately recognised. By the 1980s, the most powerful versions, with 400hp engines, could cut and load an astonishing 800 tons (813 tonnes) of coal per hour. The machine was rapidly adopted to the extent that it became the dominant tool of coal-getting by a wide margin. By 1977, an astonishing 80 per cent of the mechanised coal output was being delivered by the Anderton machines. It was assisted in this market dominance by several stages of improvement. Initially, the machines had just one fixed drum and could cut in one direction only. Then, in 1963, bi-directional shearer-loaders were introduced, and in 1969 a further step forwards was taken with the ranger-drum shearer, which had the cutter drum mounted on the end of an arm that could be raised or lowered according to the undulations of the coal seam. With the introduction of two-drum ranging machines in the second half of the 1970s, the Anderton shearer-loaders reached their optimal productive capacity. (Placing the shearer on a movable arm also meant that the machine could be used to cut extensions of the gate at the end of the runs, swinging up and over to form the arc of the tunnel.)

No coal-getting machine is perfect, and the Anderton loader was no exception. Although it was certainly efficient in cutting, the depth of the cut (known as 'the width of the web') was actually quite shallow compared to other methods of coal-getting. The width of the web for the shearer-loaders was about 21–27in (55–70cm), the latter only possible in the 1970s when stronger and longer hydraulic supports gave more space and support at the coalface. Shearer-loaders also produced a high degree of airborne coal dust, requiring intensive dust suppression systems (see Chapter 5). The grinding action of the rotary drum was also quite destructive and produced a large volume of small coal. In another age, this fact would have made the shearer-cutter non-viable. During the post-war period, however, this was not as much of a problem because small coal was actually ideally suited to being burned in coal-fired power stations.

⬆ **A side view of the Eickhoff shearer-loader system, showing how the power unit locks into the armoured flexible conveyor.** *(Author/Big Pit NCM)*

⬇ **A close-up of a shearer-drum in action. Note the angle of the cutting head, which served to direct to the cut coal onto the conveyor system.** *(Author/Cornwell/Big Pit NCM)*

➡ **A shearer-loader in action. This photograph indicates how mechanisation dramatically improved miner safety; see the distance of the miners from the cutting face and the substantial roof support.** *(Author/Big Pit NCM)*

TREPANNER

The Anderton shearer-loader was not the only new cutter type to emerge during the 1950s. A competitor was known as the trepanner, not least because its operation resembled, with some leap of the imagination, the medical process of trepanning – cutting out sections of bone from the skull of a patient. This machine, which was developed by the Anderson Boyes company, was introduced into service in 1952 and was specifically designed to cut coal in a less destructive manner than the Anderton shearer. It featured a rotary hollow-cylinder cutting wheel, typically 34in (85cm) in diameter, the cutting teeth arranged around the 'mouth' of the cylinder. When driven across the coal seam, the trepanner sliced out a large core of coal, the hollow cylinder meaning that sizeable chunks of coal were preserved and moved on to the conveyor. Because the round cutting head left large sections of coal above and below the path of the cut, the machine was also fitted with separate jibs for floor cutting, roof cutting and backshearing.

The trepanner was an effective machine, attested to by the fact that 20 per cent of mechanised coal output was won by trepanners between 1961 and 1970. Like the Anderton shearer-loader, the trepanner also started out with a one-directional cutting head, but soon acquired two (one at each end) for operation in both directions. It was also mounted directly to the AFC. An interesting development came during the 1960s, when the trepanner was combined with a shearing drum, something of a fusion between the trepanner and the Anderton. These complicated machines, however, never entirely fulfilled their productivity promises, and by the late 1970s they were completely withdrawn from the UK fields.

⬇ **A coal trepanner in action, identifiable by its distinctive longitudinal boring head.** *(Author/Big Pit NCM)*

TUNNELLING MACHINES

The mechanisation of tunnelling machines was just as important as the mechanisation of coalface operations. Access to a coal seam was only possible via roadways, gateways and other tunnels, and the speed at which they could be dug was therefore a limiting factor on a mine's productivity. A 19th-century estimate calculated that for every 984 tons (1,000 tonnes) of coal extracted, 16ft (5m) of new tunnels had to be dug (Ashworth 1986: 88). The challenge of tunnelling also included ongoing maintenance. Long and high mine tunnels are subjected to extreme multi-directional pressures, leading to convergence. Thus, regular maintenance processes included: ripping – restoring the height of a sagging roof by cutting off material; and dinting – removing stone from a rising floor.

One of the chief methods of tunnelling from the second half of the 19th century was drilling and shotfiring; indeed, even by the 1980s, 45 per cent of mine tunnelling activities were performed in this way, although mainly for smaller tunnels (Ashworth 1986: 90). Following nationalisation in 1947, however, the quest for more effective mechanised methods of tunnelling intensified.

THE CONTINUOUS MINER

The first steps forwards revolved around the mechanical loading of stone cut during the tunnelling process, achieved through loading vehicles and conveyors, but by the mid-1950s new generations of tunnelling machines were also in use. These included the Continuous Miner, a lengthy but narrow machine driven by crawlers, fitted at the front end with multiple endless-chain rippers running on a vertical plane around a ripper bar 30in (76.2cm) wide, and which could be swivelled to a 45-degree

A rotary boom-arm roadheader machine cuts into rock and coal during the development of a new mine roadway. *(Author/Big Pit NCM)*

⬆ **A continuous miner machine, with a combined cutter and conveyor.** *(Author/Mason)*

angle on both sides, and raised from floor level to roof level. The ripping head cut the stone in front of it, transferred it on the chains to a hopper just behind the head, which in turn lifted the stone on to a short length of conveyor. This conveyor dropped the stone into a central hopper, transferring the coal onwards on to a long rear conveyor, articulated so that it could swing through a horizontal arc of 90 degrees. In terms of performance, the Continuous Miner could drive a heading 11ft (3.35m) wide through a 7ft (2.1m)-high coal seam at rates of up to 345ft (105m) per shift, although these figures are evidently for driving a tunnel directly into coal, which is far easier than cutting through hard rock types.

A similar machine was the Colmol mining machine, which had a series of revolving chipping heads set in bulldozer-like blades, and included a floor-shearing blade and a roof-shearing blade. Again, it was propelled by caterpillar tracks and had its own integral conveyor system mounted behind the cutting head. The *Deputy's Manual* of 1956 notes that 'It is claimed that it will drive a heading of 4 ft [1.2m] high by 9 ft 6 in [2.8m] wide at a rate of 3 ft [0.9m] per min' (Mason 1956: 512); I have been unable to find any subsequent information confirming or questioning these figures.

⬇ **A fitter inspects the cutting teeth of a rotary roadheader machine. These were used for driving roads into coal and most types of rock, although compressed air percussive hammer drills were generally used for hard rock.** *(Author/Cornwell/Big Pit NCM)*

RIPPING MACHINES AND ROADHEADERS

The biggest advances in tunnelling, however, arrived in the 1960s and 1970s. From the early 1960s, large ripping machines steadily came into use, and with their improvement and development these began to replace blasting as the primary means of tunnelling (not just for improving roof height), although only against weaker types of rock.

There were two types of ripper. The first was the 'woodpecker' type, which had a large cutting chisel driven into the rock in a similar manner to that of a pneumatic hammer drill. These were seen either mounted on their own conveyors or suspended from the roof girders by a mounting apparatus. The more significant type of ripping machine was the rotary cutting-head 'roadheader', especially as developed by Dosco Overseas Engineering Ltd in the 1970s. The entire machine was self-propelled on caterpillar tracks, the rotary cutting-head projected forwards via a telescopic boom while other hydraulic mechanisms lifted or raised it, giving the operator a full range of movement in the cutting operation. A loading apron just beneath the cutting head scooped the broken rock from the floor and pushed it to the rear of the machine via a conveyor, where a loading jib deposited the stone on to a further conveyor or into another form of haulage to be taken away.

Roadheaders, of increasing sophistication and variety, came to be the driving force of coal-mine tunnelling; in 1981–82 they accounted for 77 per cent of the total distance of tunnels driven (Ashworth 1986: 91). Full-face tunnelling machines – machines with cutting faces of the same dimensions as the finished tunnel – were also utilised. These actually had an ancestry dating back to the 1880s, and early attempts to drive a Channel Tunnel, and they were used subsequently in various civil engineering jobs. However, the need to make them flameproof for underground operations, their extreme noise emissions plus the weakening effects they could have on the rock strata limited their use in coal mining, although some powerful machines were in use during the 1970s.

COAL PROCESSING

A study of mining mechanisation would be incomplete without expanding on what processes the coal underwent once it had reached the surface of the pit. During the days of manual coal extraction, the coal that emerged from the shaft was often quite 'clean', in the sense that it contained relatively few non-coal minerals, according to the precision of the miner's pick. In the age of mechanised cutting, however, as much as 40 per cent of the rock cut might actually be shale. This, and the fact that the coal itself varied in size of pieces (with different markets for different sizes), meant that the coal traditionally had to go through three stages before it could head out of the colliery for sale or use: separating, picking and washing.

Before looking at those three processes, however, we should note what actually

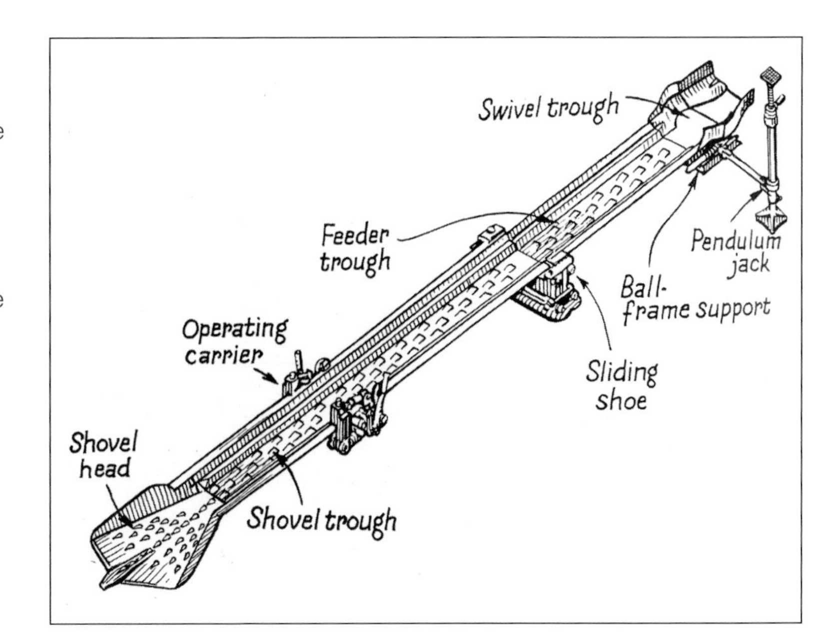

⬆ **A duckbill loader, gathering the coal through a reciprocating shovel.** *(Author/Mason)*

⬇ **A tub creeper mechanism, used for retarding the descent of tubs.** *(Author/Boulton)*

➡ **A tub tippler. Note how the tub has been completely upended in the rotating mechanism.** *(Author/Big Pit NCM)*

➡➡ **A late 19th-century scene of coal screening, the 'pickers' removing pieces of 'dirt' (i.e. anything that was not coal).** *(Author/Tomalin)*

⬇ **A close-up of the water-balancing headgear, an early form of headgear.** *(Author/Big Pit NCM)*

happened to the tubs of coal once they emerged from the shaft. Based on best practice by the early 20th century, the full tubs would be pushed from the cage on to a localised rail network. If the tub were filled with waste products only (stone, dust etc.), points on the rail tracks would divert the tub off to a spoil tip, or slag heap. Tubs filled with usable coal went to a weighing station. Here, the tub would be driven on to a weigh bridge, and the overall weight was displayed on a dial. Once this had been recorded, the tub was then advanced to a tippler – a mechanism that would empty the tub of coal. There were numerous types of tipplers, from gravity-operated to fully powered varieties, but they worked largely with the same purpose: holding the tub firmly in a frame, then rotating the frame so that the coal was tipped out down a chute for passage to the next stage of its journey.

While, in the pre-nationalisation days, coal was generally transported around the pit surface in various cars and wheeled vehicles, the second half of the 20th century saw the rise of more advanced conveyor-based coal haulage systems on the surface. Some mines developed sophisticated coal preparation plants on site. These comprised a dedicated group of buildings devoted to all aspects of readying the coal for market, including screening, washing and crushing (when larger

pieces of coal had to be crushed down to a smaller size for a specific industry). These plants might have literally miles of conveyors taking the coal from one station or 'transfer house' (much like an underground loading station, but on the surface) to the next, the conveyors and their contents protected by distinctive metal frame housings.

The Distl-Susky Screen was a roller-type screen for treating large pieces of coal. One screen could handle 200 tons of coal per hour. *(Author/Big Pit NCM)*

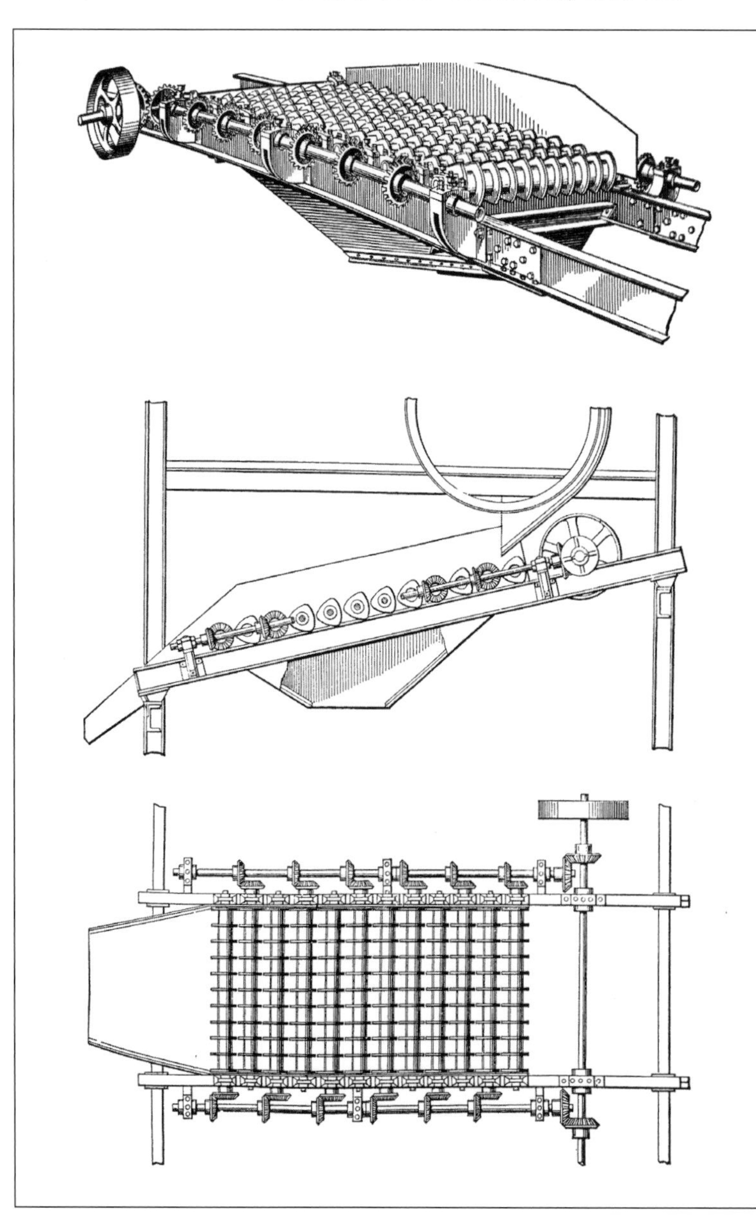

SEPARATING

The separating process was developed to sort the run-on-mine (ROM) coal – the coal as it came direct from the coalface – into its different sizes. The first iteration of screening was simply rows of stationary fixed steel bars, inclined at an angle of 23–30 degrees, through which the coal either fell or was trapped depending on its size, but by the early 20th century several mechanised types were in use. These broke down into the following types:

- **Moving-bar screens** – Screens of bars, moving in relation to one another, separated out the different-sized coals.
- **Roller screens** – The ROM coal moved over a sequence of powered rollers, with the small coal falling down through the gaps between the rollers.
- **Rocking screens** – The screening surface was composed of hinged sections moving in a rocking action, shifting the coal forwards while at the same time separating out the different sizes.
- **Shaking/jigging screens** – This type had a large screen fixed in a frame, which was moved in a backward-and-forward motion, with the screen slightly inclined to move the coal forwards to the next stage of processing.
- **Multi-motion screen** – This screen was moved in both up-and-down and forward-and-backward directions.
- **Gyrating screen** – Here, the screen movement described a gyrating or circular pattern in the horizontal plane.
- **Revolving-drum screen** – This separating method used concentric screening drums (typically three), the coal being shaken through the screens while also, by virtue of the overall cone-shaped profile of the drums, being advanced forwards to an exit point.

PICKING

Picking referred to the activity of removing non-coal materials, such as shale, stone or clay, from the ROM coal. The simplest way of doing this, and one that was applied for centuries, was hand-picking. Here, human 'pickers' would visually identify the rock to be removed, and did so by hand, working either from a travelling picking band – a conveyor, which might also be vibrating or oscillating to encourage separation, typically moving at speeds of 30–60ft (9–18m) per minute – or a picking

An early 20th-century scene of manual coal screening, the 'pickers' categorising the coal by size. *(Author/Big Pit NCM)*

table, which could be either straight or circular and revolving.

WASHING

Washing the coal was typically the final stage of the coal preparation. Its objective was 'the purification of the coal by removing the admixed particles of stone, shale, or pyrites. It is generally applied to the sizes which are too small to be economically cleaned by hand-picking' (Boulton 1908: 330).

Washing was, and remains, a highly technical process, with many different approaches. Its core principle is that coal and shale have differences in their specific gravity, which means that when they are suspended or moved in a medium such as water – the most common medium for washing – or even air, the valuable, lighter coal would be separated out from the heavier shale. Coal washing was introduced in *c.* 1830, and by the early 20th century coal-washing plants were major industrial units of great sophistication, containing air-blower separators, water-washing machines, separating screens, dryers (for drying the wet coal) and many other mechanisms.

One of the most sophisticated washing processes, introduced *c.* 1906, was froth flotation. In this system, the coal dust, or ROM crushed into fine particles, was placed into a tank of oily water. Small air bubbles were then flowed through the water, and these attached themselves to the lighter coal dust, carrying it to the surface, where it could be skimmed off, while the heavier shale fell to the bottom. Froth flotation became particularly sophisticated in the second half of the 20th century with the refinement of a range of chemicals – known as frothers, collectors and depressants – which improved the speeds, volumes and quality outcomes of the washing process.

COMPUTERISATION AND AUTOMATION

The 1960s was a time in which the world steadily woke up to the potential of advanced electronic technology for automating processes or performing them by remote control, and the coal-mining industry was no exception. One of the first visions in this regard was the Remotely Operated Longwall Face (ROLF) concept. This attempted to realise a mining process in which most of the vital engineering activities were controlled electronically from a central point distant from the coalface itself. First trials of the system began in 1962, but as the 1960s progressed the project was steadily abandoned owing

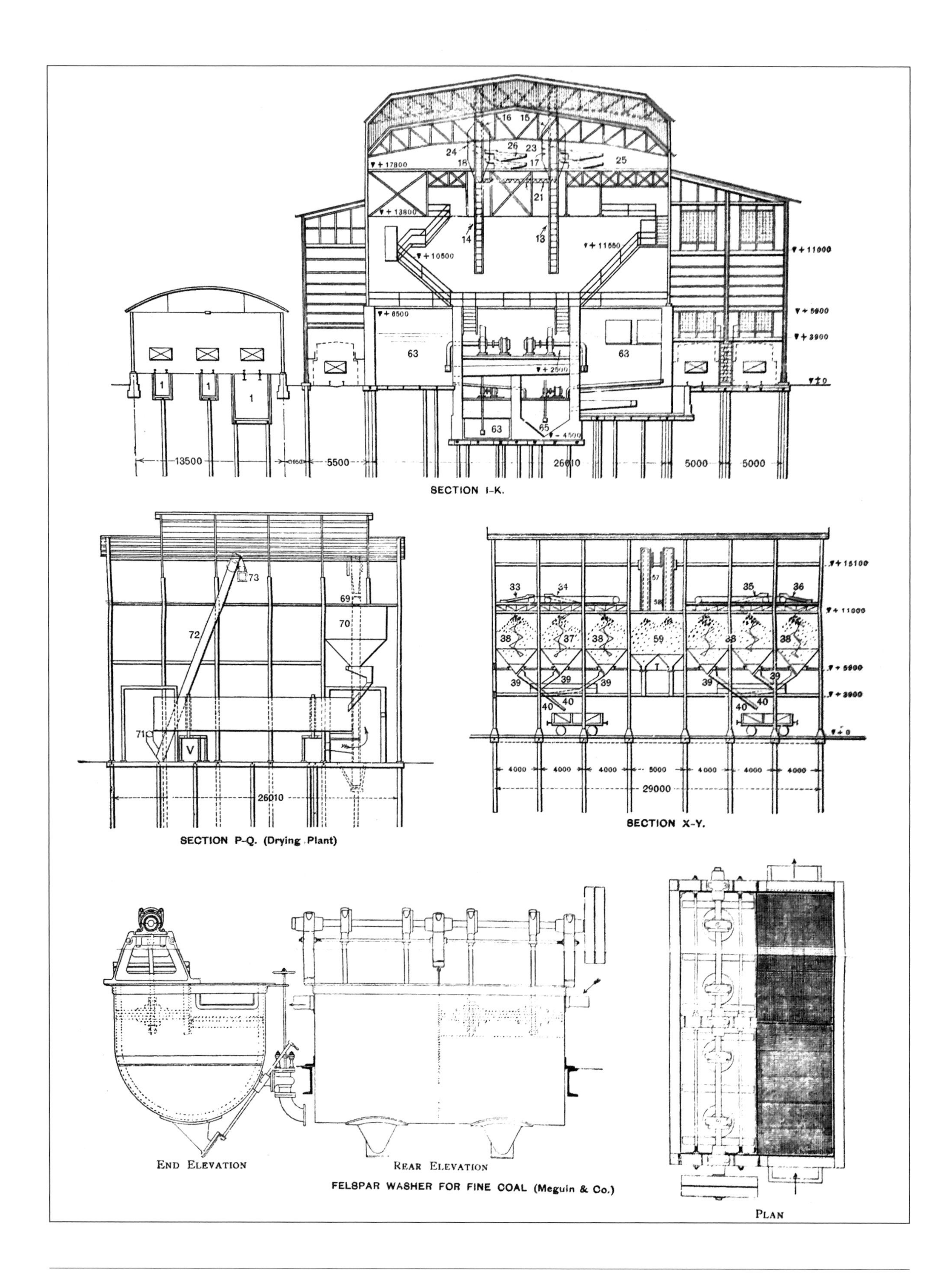

FELSPAR WASHER FOR FINE COAL (Meguin & Co.)

⬅ **An impressive diagram from 1908 showing a coal-washing plant. The Felspar Washer used pulsating water to separate the particles.** *(Author/Boulton)*

to the technological limits of remote control at that time, although much useful research information was gleaned. Automation, rather than remote control, was managing to establish itself, however, particularly in the fields of haulage, winding and hydraulic prop control.

The next step towards realising the benefits of hi-tech systems came in 1974, with the concept of Advanced Technology Mining (ATM), supported by work done at the NCB's Mining Research and Development Establishment (MRDE) at Bretby, near Burton upon Trent. One of the products of the research was to improve the system by which a bank of hydraulic props could be advanced by a single operator, significantly reducing the number of men required to prepare the coalface after cutting. Another ingenious device, riding on the increasing miniaturisation of computer components, was the nucleonic head. This device was fitted to shearer-cutters, and it automatically analysed, through the projection of cobalt 60 particles, the different gamma radiation emissions from the rock, which enabled the nucleonic head to tell the difference between coal and other rock types. This information was fed to a processor that automatically adjusted the path of the shearer to avoid the rock and concentrate on the coal.

Although the original vision of fully automated and remotely controlled coal-mining operations was never fully realised during the lifetime of the British mining industry, computerisation certainly transformed the industry during the 1970s and 1980s, as it did the rest of the world. As computing power grew exponentially, computer systems were used extensively to form networks of monitoring and control systems. In the late 1970s and early 1980s, many collieries adopted specific sets of standardised mining computer programs, particularly the Mine Operating System (MINOS) and the Face Information Digested On-line (FIDO). These controlled or monitored multiple aspects of the mine operation – including ventilation and monitoring of airflow, storage and movement of coal, pumping systems and drainage, conveyor control, and the movement of inventory of equipment – from a centralised control room.

⬆ **Computerisation during the 1970s and 80s led to the automation of many surface processes.** *(Author/Big Pit NCM)*

* * *

Through intensive modernisation in the 20th century, the British coal mining industry managed to reinvent itself several times. By the 1990s, the practicalities and productivity of coal mining were virtually unrecognisable compared to 150 years previously. Yet some of the dangers of mining remained an unwelcome constant, and it is these we will explore in the next chapter.

⬇ **This image gives a good impression of how the mine tubs were moved between the shafts and the surface workings.** *(Author/Big Pit NCM)*

MINE SAFETY: VENTILATION, DRAINAGE, SUPPORTS, DUST AND FIRE

For centuries, coal mining was one of the most dangerous occupations on the planet, with literally thousands of miners being killed globally, and hundreds in Britain, every year. Each mine shared a similar group of challenges, to varying degrees, in the quest to make the mine a viable space for production and a safe space for its workers.

← Powered hydraulic supports such as those seen here helped to make deaths from roof collapses relatively rare, although the presence of powerful machinery and unpredictable geology means that mining remains a dangerous occupation. *(Author/Big Pit NCM)*

VENTILATION

The need for proper ventilation of any small space seems self-evident, and is usually easily achieved. In a coal mine, however, ventilation takes on a whole new level of complexity and importance. Not only is it a matter of getting adequate air supplies down to men who might literally be several miles from the source of air ingress, but it also has a critical part to play in preventing the build-up of poisonous gases and in reducing the potential for underground explosions.

AIR QUALITY AND 'DAMPS'

For miners to work safely in a breathable environment, the oxygen content of the air should not be less than 19 per cent. Deep beneath the ground, however, there are several factors that can affect this balance:

- **Respiration** – The physical activity of humans breathing decreases oxygen content and raises carbon dioxide levels in a confined space, if there is no or little input of fresh air.
- **Machinery** – Various machines affect proximate air quality, either through consuming air as part of their working process or through emitting gases such as carbon monoxide and carbon dioxide as by-products.
- **Ground emissions** – Coal releases methane and a host of other gases as soon as it is cut. These gases can affect both the breathability of the air and, in the case of methane, create the conditions for an atmospheric explosion.

In mining terminology, the word 'damp' – attached to a qualifying prefix – defined a range of dangerous air conditions. The most serious of all was 'firedamp', which was principally methane. Firedamp is released in significant volumes from cut bituminous coal, and if present in the air in proportions of 5–14 per cent it becomes flammable. It was typically the catalyst for the many devastating mining explosions that took the lives of thousands of miners. The insidious problem of firedamp for miners is that it is odourless, and even at the concentrations described above, it is breathable. Therefore, technologies are required for its detection.

In addition to firedamp, there were four other problematic damps in British mines:

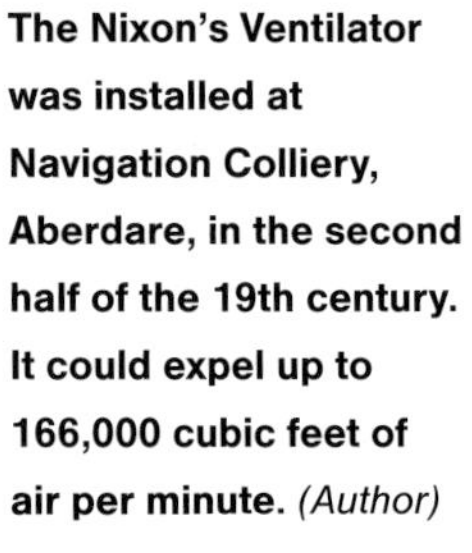

The Nixon's Ventilator was installed at Navigation Colliery, Aberdare, in the second half of the 19th century. It could expel up to 166,000 cubic feet of air per minute. *(Author)*

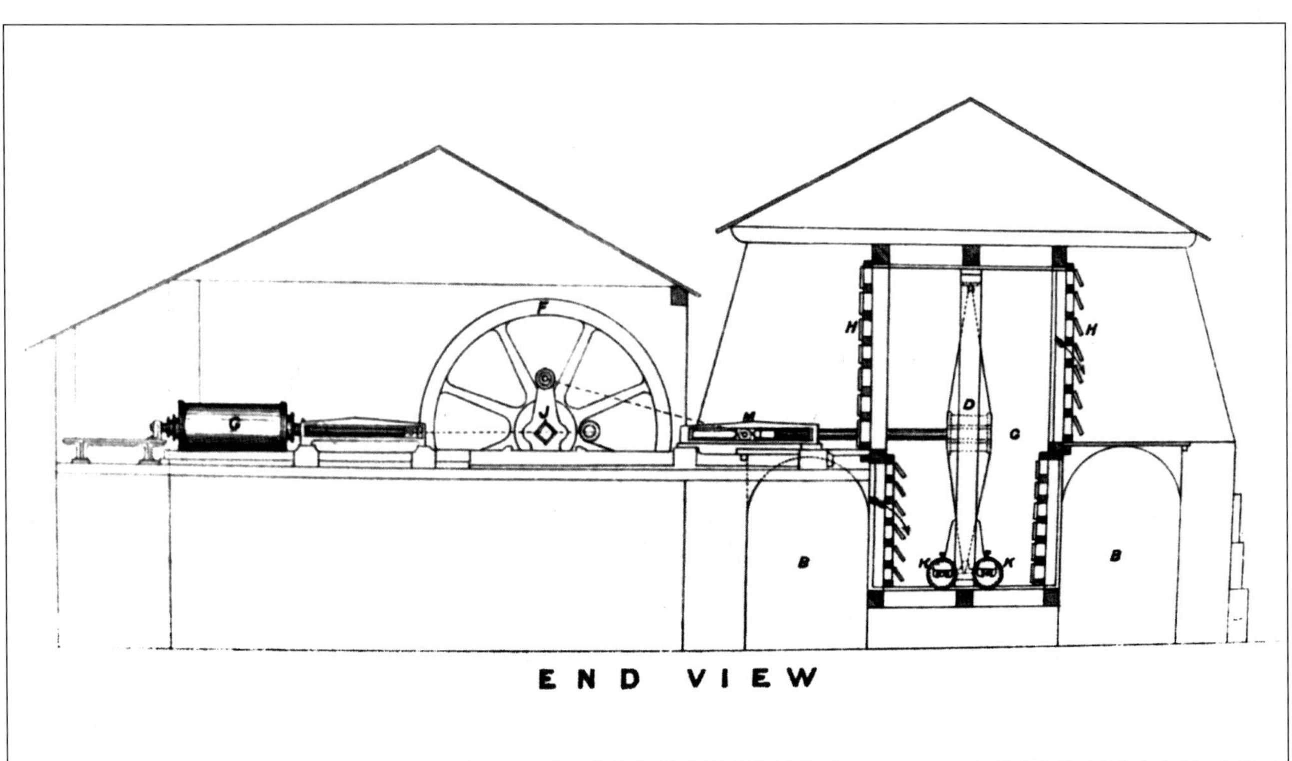

- **Blackdamp** – This heavy gas is composed of carbon dioxide and nitrogen. Because it is heavier than air, it was particularly dangerous to miners in low workings; elevated levels could result in unconsciousness, suffocation and death.
- **Whitedamp** – This is carbon monoxide, produced mostly by combustion processes such as fires and engines. In mines, it tended to occur after fires and explosions, including after shotfiring. Like methane, carbon monoxide is odourless but also poisonous (it prevents the blood from bringing oxygen to cells, tissues, the brain and organs) and in high concentrations, lethal.
- **Afterdamp** – This is the name for the mixture of gases found in the air of a mine after an explosion, of varying degrees of threat to humans. The most dangerous of them is carbon monoxide, but burning materials could also produce a cocktail of poisonous airborne chemicals.
- **Stinkdamp** – This term refers specifically to hydrogen sulphide. A poisonous gas, stinkdamp had a very definable odour, like that of rotten eggs, so unlike the others listed above was detectable by the human nose.

Bearing in mind the list of noxious and dangerous gases present in a coal mine, one of the key purposes of ventilation was to keep the air moving, so that the gases were constantly pushed on and through the workings, replaced by fresh air and eventually driven out of the mine.

AIRFLOW

The key objective in mine ventilation is providing a constant flow of fresh air through all of the tunnels. Achieving this in Britain's deep mines involved detailed planning. (Each mine would literally produce an air plan, mapping out the way that air flowed through the works.) Simply having a single open shaft would not generate this flow, especially in deep mines with, perhaps, 20–30 miles (32–48km) of tunnels below. In such cases, the air would not flow, but rather merely settle around the base areas of the shaft. This was one of the reasons that deep mines came to have two shafts (or drifts), separated a

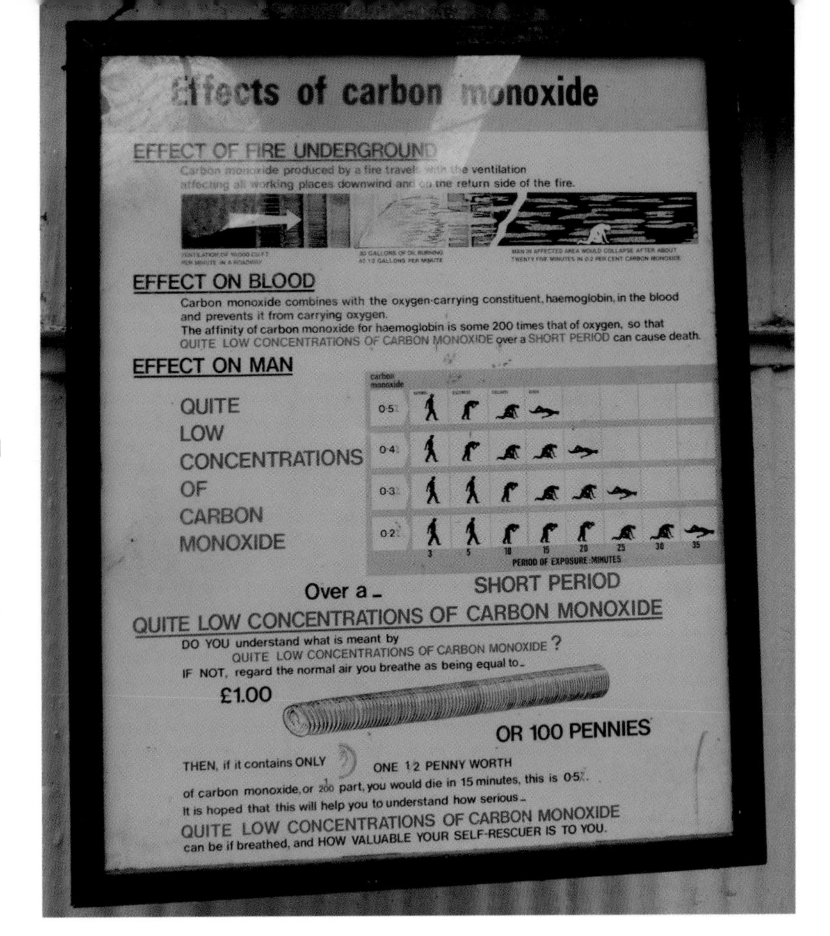

⬆ **A mine poster explaining the effects and dangers of carbon monoxide poisoning.** *(Author/Big Pit NCM)*

⬇ **Mine ventilation via bellows from the 16th-century German manual *De re metallica* by Agricola, a pseudonym of one Georg Bauer.** *(Author/Big Pit NCM)* .

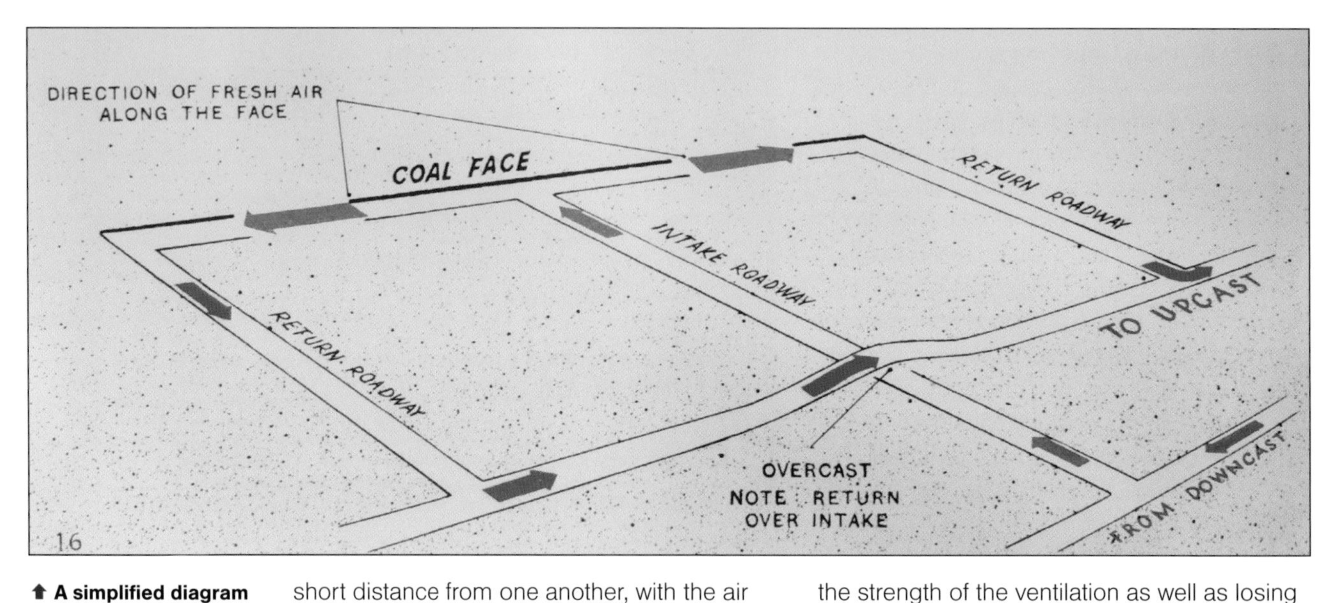

⬆ **A simplified diagram illustrating the flow of air around a mine, the blue arrows showing fresh air from the downcast, the red arrows stale air to the upcast.** *(NCB)*

short distance from one another, with the air pressure differentials created between the two creating a natural movement of ventilation. One shaft functioned as the 'downcast' – the shaft down which fresh air was drawn into the mine; and the other as the 'upcast' – the shaft through which the circulated air exited. Note that legislation eventually stipulated that every mine had to have two routes of entry and exit. Some mines that had an alternative route of exit might instead have a pure ventilation shaft, rather than an active winding shaft, just for the purposes of upcast.

Natural ventilation was not an ideal solution, as the airflow weakened as it wound its way around extensive mines. The air might also go into non-working areas, again reducing the strength of the ventilation as well as losing valuable airflow to non-populated areas. During the 1760s, therefore, the system of 'coursing the air' was developed. Special ventilation doors, or 'traps', were installed at key junctures between gates, roadways and coalfaces, directing the airflow as desired. Individuals known as 'trappers' – for many years children, hunched for long hours in intermittent absolute darkness (they usually had candles, but sometimes extinguished them for periods to conserve candle wax) – manned the doors, opening them only for the passage of corves or tubs. The trappers would eventually, mercifully, disappear under the corrective of child labour laws, and were replaced by self-closing and automated doors. At some points, 'regulators' might also be installed. These were heavy steel doors that featured a large sliding panel (rather like the inspection portal on a prison door), which could be locked in place in various degrees of openness. This meant that while the main part of the door would still perform its function of directing the flow of air through the principal workings, the open portal could direct a portion of the airflow into other parts of the mine, as required. When a tunnel was no longer in use (usually because it went to a part of the coalface that had been worked out), it was generally blocked off permanently, in a process called 'bashing up'. Here, the tunnel was sealed with a thick wall of stone or brick, sometimes up to 65ft (20m) thick. An important consideration with bashing up was that on the other side of the

⬇ **A ventilation duct leading out from the fan house at Big Pit National Coal Museum.** *(Author/Big Pit NCM)*

wall, methane and toxic gases would continue to build up. This was a concern, because in an active mine there was no guarantee that the tunnels wouldn't intentionally be reopened one day, or at least accidentally penetrated. For this reason, the sealing wall might sometimes feature a tapped ventilation pipe running through it, to drain off the gas when it was safe to do so. (As a side point, abandoned mine workings could also fill up with water, and if at a later date the tunnel was penetrated, such as through roof bolting, there could be a terrifying inrush of water, or a potentially lethal blast of pressurised gas containing fragments of rock.)

FIRE AND FURNACE VENTILATION

Air coursing improved the flow of air around the mine, but what was still needed was powered ventilation, which required physically controlling the dynamics of the airflow. The first method of doing so, recorded in the 1750s, was fire baskets. These were little more than burning braziers suspended over the upcast. In the same way that an open domestic fire can generate a forceful draught across the room to sustain it, as air rushes in to fill the space left by the rising heated air, the fire basket produced stronger ventilation airflow around the mine through increasing the low pressure in the upcast. Later, underground furnaces were also installed, but these brought with them natural fire hazards, resulting in some explosions. A countermeasure to this danger was to create a dumb drift in which the fire was located, which meant that any contaminated or flammable air flowing into the upcast did not pass over the flames.

The furnace system of ventilation was employed throughout the 19th century and into the 20th century, and could generate considerable airflows; in 1925, one such system in Walsall Wood Colliery in South Staffordshire was pulling through 100,000 cubic feet (2,832 cubic metres) of air every minute. But given the safety considerations that hung over furnace ventilation, engineers of the 19th century also began to develop safer and more efficient mechanical ventilation systems.

STEAM-POWERED VENTILATION

The first of these experiments took place in the late 18th century, with John Smeaton's reciprocating steam pump system, designed for extracting the foul air from mine tunnels. The next step came in 1807, with John Buddle's installation of a steam-driven air pump at Hebburn Colliery on Tyneside. The machine evacuated about 6,000 cubic feet (170 cubic metres) of air a minute from the upcast shaft top; this was actually well below what was required, as a 19th-century pit typically demanded an air circulation of 30,000–50,000 cubic feet (850–1,415 cubic metres) per minute. A far more effective steam

The compound Corliss Riedler pump was one of several types of Riedler pump introduced in the late 19th century. It was highly efficient, pumping up to 2,000 gallons of water per minute. *(Author/Boulton)*

⬆ **A powerful two-stage, twin-cylinder, double-acting and water-cooled air compressor unit in a colliery fan house.** *(Author/Big Pit NCM)*

ventilation system was developed by William Price Struve of Swansea, who used two large cylindrical pistons, each working as a double-acting pump, to deliver an efficiency more in the right ballpark: it circulated 56,000 cubic feet (1,585 cubic metres) of air per minute.

FANS

The most important of the ventilation developments in the 19th century were steam-powered centrifugal fans, the first of which was installed in service in 1836 at Hemfield pit in South Yorkshire. Over the rest of the century, fans progressively took over mine ventilation, being effective, reliable and mechanically simpler than many of the steam pump types. The fan diameters generally ranged between 9ft and 45ft (2.7m and 13.7m). Driven by steam, they would often run at about 60–70rpm. By the mid-20th century, the efficiency of such ventilation fans, now electrically driven, had made giant strides. For example, a Sirocco fan had a double inlet, each with 64 blades, a diameter of 13ft 6in (4m), but could run at 160rpm, which delivered 494,000 cubic feet (13,990 cubic metres) of airflow per minute.

Another fan utilised in the modern British mines was the axial flow type, mainly applied to ventilating headings and drifts: they were typically mounted on the roofs of the tunnels, rather than on the surface. These compact units drew in and pushed out air in a similar manner to the propeller of an aircraft or the screw of a ship. Although these units were noisy, they gave excellent outputs.

⬇ **A high-speed Walker fan used to deliver underground ventilation, installed c. 1910.** *(Author/Big Pit NCM)*

SELECT KEY POINTS OF VENTILATION PLANNING IN DEEP MINES

- Haulage roads should be kept as intakes.
- When supplying air to inclined seams, it is best to supply the air to the lowest point in the district (a designated working area of the mine) first and take it from the highest place last.
- Areas that have prominent faults are usually prone to gas build-up, and therefore should be ventilated last to avoid distributing the dangerous gases around long expanses of workings.
- Every road that does not require ventilation should be sealed.
- Doors and regulators should be kept free from haulage routes as far as possible.
- Air measurement and sampling stations should be established at regular points to measure the velocity and the quality of the airflow.
- The direction of airflow should be marked clearly on signs on the walls, especially at the point of splits.

⬅ **Two booster fans at the end of a mine roadway. Brick packs, such as that behind the fans, had to be sealed as tight as possible to prevent air leakages.** *(Author/Big Pit NCM)*

⬇ **A cross-section of a fan house, illustrating how the fan drift connected out to the upcast shaft.** *(Author/Big Pit NCM)*

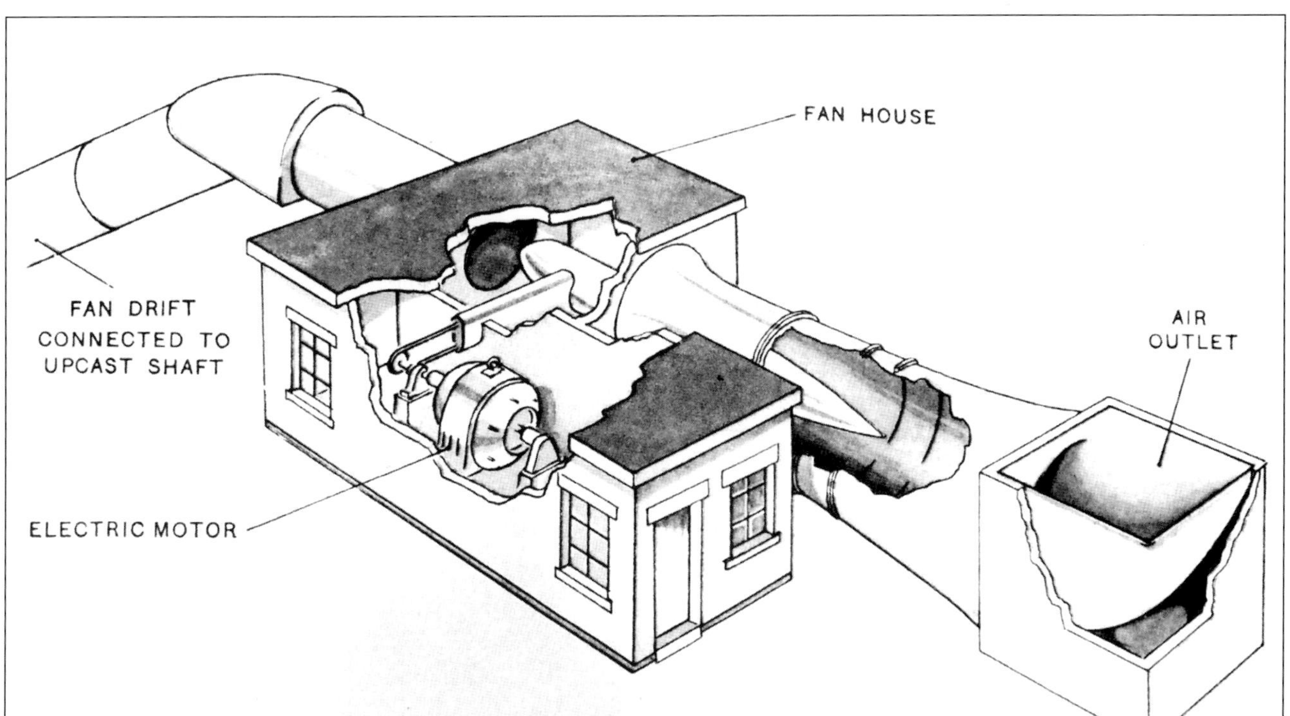

One of the larger types, made by Walker-Macard, could run up to 280rpm with its 16ft 6in (5m) rotor and generate airflow of 600,000 cubic feet (16,990 cubic metres) per minute. It was through such means, accompanied by constant mechanical tweaks and improvements, that mine ventilation was maintained through to the closure of Britain's last deep mine.

EXPLOSIONS

Mining explosions could be of a devastating magnitude. The worst of all, indeed the worst mining disaster in British history, was that which occurred at Senghenydd in 1913. The lethality of the event had many contributing factors, not least the suffocation of dozens of men in the aftermath of the explosion, as the mine workings were stripped of air and filled with poisonous gases. But this disaster, and dozens of lesser but highly costly others, were the bane of colliery life for generations of miners and their families, and remain a threat to this day. As an example of the frequency of colliery explosions, there were a total of 643 significant detonations in underground mines between 1835 and 1850 alone. Between 1870 and 1880, 2,700 miners lost their lives to explosions. To remind us of the continuing and present danger, in the USA there were 10 multiple fatality explosions in underground coal mines from 1986 to 2010.

There were three main causes of mining fires and explosions. First, the ignition of firedamp, in the concentrations described above (the perfect ratio for explosive combustion is 9.5 per cent), when the methane gas came into contact with an open flame, heat source or spark. Second, coal dust lying on surfaces catching fire under open flame or through contact with hot machinery, although the ignition of coal dust lying inert on the floor rarely occurred, and could only be achieved under the intense heat of a concentrated direct flame. The third, and most dangerous, reason, however, was airborne coal dust. Coal dust is highly combustible when it is suspended as a dense cloud of particles in the air. The explosion of dust generated by the process of coal cutting was a flammable threat, but the real danger came from the interaction between a methane explosion and coal dust. In the most serious explosions, the initial flash point was typically a localised methane explosion, the shock wave of which lifted the coal dust from the surrounding surfaces. This material then exploded itself, the dust acting as fuel for the methane blast (to look at it another way, the methane essentially acts as a detonator for the coal dust). There began a truly vicious cycle, each blast lifting more coal dust, which then ignites and extends the explosion, the blast running faster and faster through the tunnels and up the shafts.

STONE-DUST BARRIERS

One ingenious means of stopping the inexorable expansion of a coal-dust explosion was the use of stone-dust barriers. The barriers were shelves held up loosely on a bracket fitted to the walls of roadways, and covered with a thick layer of stone dust. This type of dust is an inert material, and has none of the combustible properties of coal dust. The barriers worked in the following way. In the event of a coal-dust explosion, the blast lifted the shelves off their brackets and threw the stone dust up into the air. This dust immediately mixed with the coal dust, blanketing the flame and stopping the further propagation of the explosion.

GAS DETECTION AND SAFE ILLUMINATION

During the 17th century, as mines became steadily deeper and therefore gassier (firedamp is rarely a problem in very shallow mines), there was an evident tension between the primary form of mine illumination – an open-flame candle or torch – and the presence of combustible methane. The candle did the miner a favour, however, in that it was recognised that an elevated light blue cap on top of the flame indicated that the candle was burning methane, and that the workings needed to be evacuated. This principle was later more scientifically captured in the miner's safety lamp (see below).

In the days before adequate ventilation cleared the gases, the only solution to making a tunnel or face safe might be to burn off the gas. This was conducted by a hapless individual known as a 'penitent' – a label given on account of his clothing of water-soaked rags and the fact that he would have to lie face down on the floor to perform his main role. This involved extending a lit candle on a long pole into the gassy space, igniting the gases in an overhead blast of flame. Needless to say, this was a job that carried with it high levels of personal risk.

During the 19th century, other methods were developed in both gas detection and management. It was noted, for example, that Monday mornings could be one of the most dangerous times for fires and explosions, because the absence of lit candles and furnaces over the men's Sunday off meant that there were higher concentrations of gas build-up by the time they returned to work. It was also observed that a fall in barometric pressure could produce greater gas emissions from the coal, through a slight increase in air pressure. Apart from ventilation, one particularly productive response to methane build-up was to drill boreholes into the coalface and tap off the methane gas to the surface. There it was often burned off, in the same manner as oil rigs burns off their excess gases, but it also became common practice to use the methane to power lamps and other surface illumination. During the 20th century, it was even sometimes channelled into colliery boilers or to nearby power stations.

Miners in the 20th century would become very familiar with the diagram below, showing how to detect firedamp from the behaviour of the safety lamp flame.

Figure 1. Testing Flame alone.

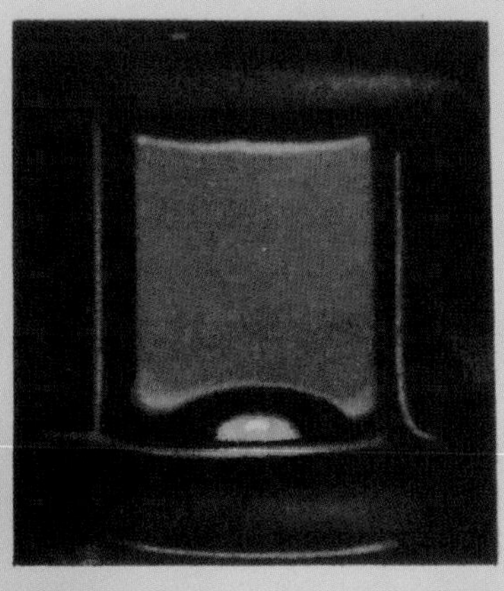

Figure 2. About 1¼ per cent. Firedamp.
Switch off the electric current.
In naked lamp mines workmen must leave the affected area.

Figure 3. About 2 per cent. Firedamp.

Figure 4. About 2½ per cent. Firedamp
All workmen must leave the affected area.

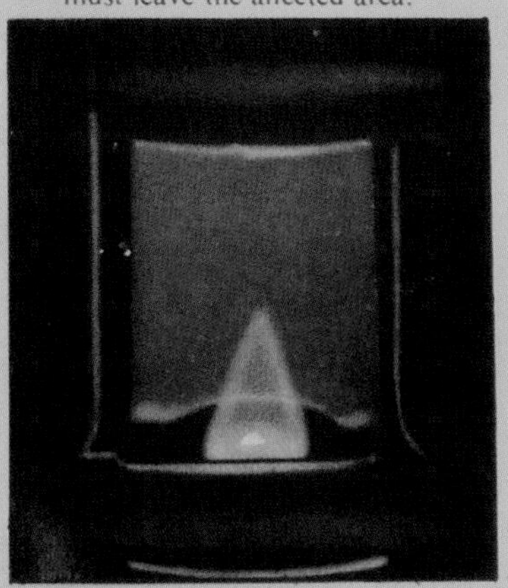

Figure 5. About 3 per cent. Firedamp.

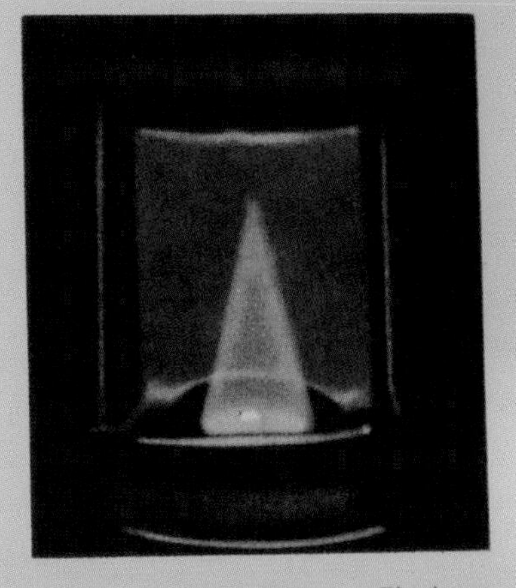

Figure 6. About 4 per cent. Firedamp.

Reproduced from the publication "Beware Firedamp" with the permission of the Controller of H.M. Stationery Office.

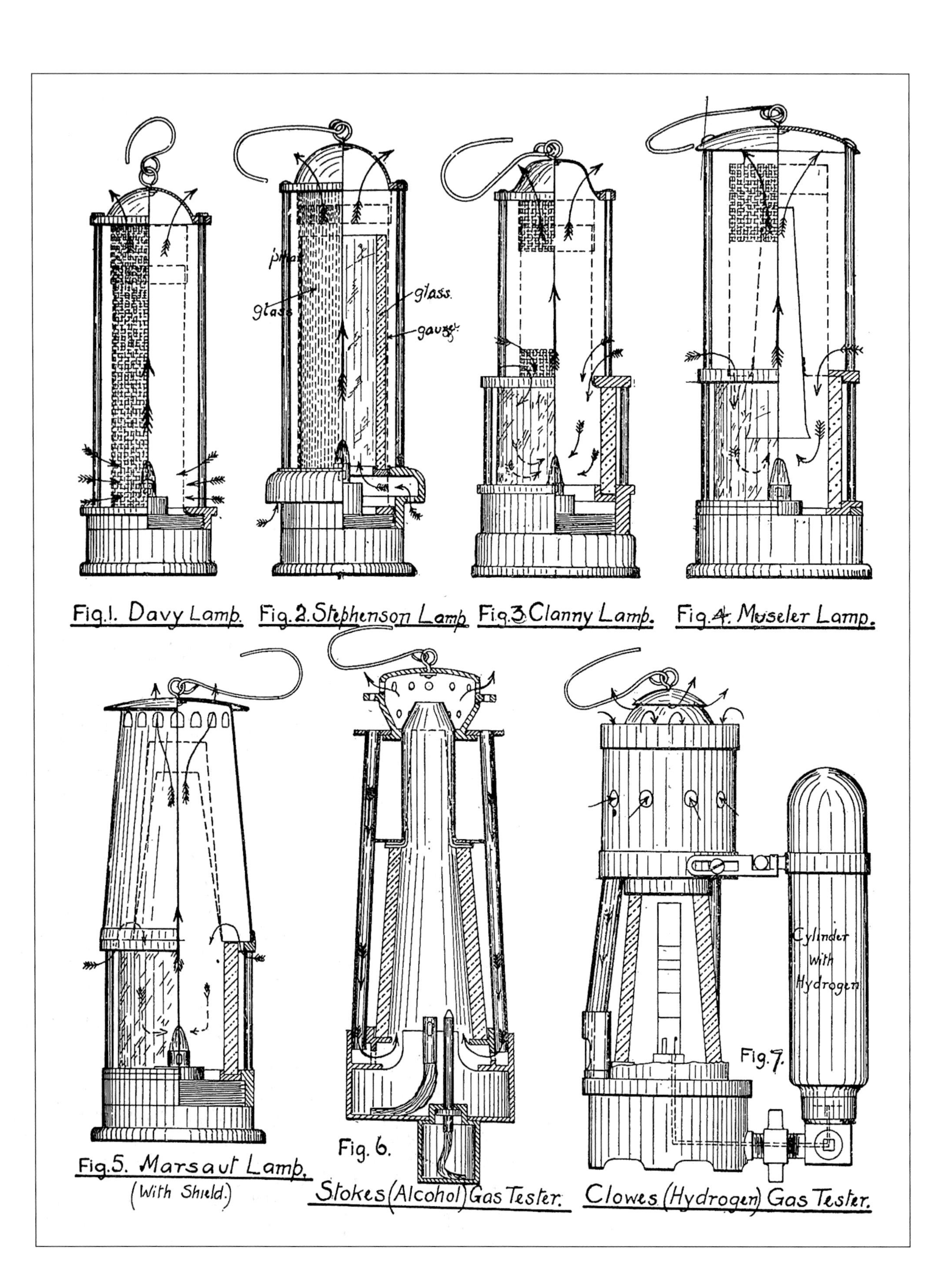

Fig. 1. Davy Lamp. Fig. 2. Stephenson Lamp. Fig. 3. Clanny Lamp. Fig. 4. Mueseler Lamp.

Fig. 5. Marsaut Lamp. (With Shield.)

Stokes (Alcohol) Gas Tester.

Clowes (Hydrogen) Gas Tester.

THE SAFETY LAMP

The matter of illumination is a pressing one in deep mining, and very much related to the topic of dangerous gases. Underground, the darkness of an unilluminated working is absolute, hence miners needed lighting that was sufficient for them to both find their way around the mine and perform their often-intricate tasks. Adding to the problem was the fact that open flames were dangerous for reasons outlined previously. Open candles and oil lights nevertheless persisted in coal mines into the 19th century, either carried by hand, mounted on wall brackets or fixed on the front of helmets. However, even before the turn of the 19th century, engineers were seeking safer means of illumination.

The first of these efforts was a strange mechanical device produced by colliery steward and mining engineer Carlisle Spedding at some point around the mid-1700s. This hand-held device featured a cogged wheel, which when turned by an operator via a handle rotated a plain steel wheel that grated against a piece of flint, producing a shower of sparks. These sparks not only provided dim illumination, but they also changed their colour according to the relative presence of firedamp, hence were an early detection technology. The sparks were ostensibly too cool to ignite the firedamp, but the 'steel mill' did actually cause an explosion at Wallsend in 1785. This fact, combined with their poor light and impracticality (a boy had to be employed specifically to operate the lamp), meant they soon went out of service.

Other methods were tried in a quest to find the ideal mining illumination, including reflecting sunlight via mirrors down the shafts and tunnels and the use of naturally phosphorescent materials and glowing chemicals, but these produced few decent results. However, following a major colliery explosion at Felling Colliery, County Durham on 25 May 1812, the race to find safe mine illumination became a national priority, encouraged partly through the newly formed Society in Sunderland for Preventing Accidents in Coal Mines. They approached the great Cornish chemist and inventor Sir Humphry Davy in 1815 to develop a new source of illumination, by which time two other inventors – William Reid Clanny and George Stephenson (of railways fame) – were already working in the field. The priority for all was to find a way of generating the flame to provide light, while shielding that flame and its associated heat from contact with the dangerous gas.

SAFETY LAMP DESIGNS

Between 1813 and 1817, Clanny offered several safety lamp types. The first of these was the Blast Lamp. This enclosed the flame in a lantern, with cisterns of water above and below the lamp to disperse heat. Air was pumped into the lamp from the bottom via a set of bellows, and hot air was vented out at the top through a chimney. Clanny went on to provide two improved versions of the lamp: the Steam Lamp in 1816 and the Gas-light Lamp in 1817. The former used a single cistern of water above the lamp to generate steam (which in turn neutralised any build-up of explosive gases inside the lamp), while the latter followed the same principles but actually ran off firedamp, not oil. These lamps were functional solutions, and were used in mines; Clanny was given several awards from the Royal Society of Arts for his inventions.

←← Six of the many dozens of varieties of safety lamps and gas testing lamps available in the first half of the 20th century. *(Author/ Williams & Cryer)*

↓ Spedding's steel mill, invented in the 1730s, indicated the presence of dangerous gases through the changing colour of its sparks from the flint and wheel, when the handle was turned. *(Anagoria/ CC BY 3.0)*

⬆ A traditional safety lamp, as used in British and international mines from the 19th century. *(Steve Pleydell/Shutterstock)*

➡ An electric hand lamp, the long steel case holding either an acid or an alkaline battery. *(Author/Big Pit NCM)*

Stephenson's contribution to the evolution of the safety lamp was even more significant. His Tube and Slider Lamp was an oil-fuelled type with the air intake, via a tube in the base, controlled by a slider on the bottom of the lamp. It also had its air tube wrapped in a section of wire gauze to disperse heat. Stephenson went on to make several improvements to the lamp. His final version, of 1815, had the flame encased in a glass cylinder and wrapped in a wire gauze bonnet, rather like the lamp that Davy would go on to create. (There was later much debate about who invented the safety lamp, but credit must surely go jointly to Davy and Stephenson.)

Davy's safety lamp appeared in its initial form in early 1816, after he devoted some time to the theoretical study of the problem. The principle behind his lamp was simple. A tube of wire gauze, made from extremely fine wire with nearly 800 tiny apertures, enclosed the flame above and below. Thus configured, air could flow into the lamp, taking firedamp with it, but the gauze dispersed the heat generated, lowering it to levels where it couldn't ignite the gases outside the lamp. (The upper part of the gauze was also doubled to cope with the higher heat emissions there.) Later, the lamp was fitted into a metal case, to reduce the risk of atmospheric detonations when strong air currents were blowing across the lamp.

The Stephenson and Davy lamps came into service shortly after their invention and spread not only through the British coalfields but also throughout the mining world. In fact, their adoption was more enthusiastic in Continental Europe than in Britain, and subsequently French and Belgian engineers made significant modifications to the lamp design, improving both the safety of the lamps and the illumination they emitted. Key improvements were the addition of a metal bonnet over double or treble gauze cylinders to protect the flame in high winds, and also the use of glass lenses to intensify the light.

SAFETY LAMPS IN USE

The safety lamps, of which 740,000 were in use in Britain by 1913, made one of the most profound contributions to mining safety. At first, their introduction seemed to have had little impact on reducing explosions; in fact, the numbers of explosions increased owing to over-confident lamp holders venturing into gassy parts of mines where no flame illumination at all was safe. But once the lamps were used with functional common sense, they were a game changer.

The lamps had two primary functions. First, they delivered a safe form of illumination. Second, they gave an indicator of the levels of methane in the air, based on the height of the blue cap atop the main flame. Posters were produced illustrating these heights and giving instructions based upon them, and some lamps even had scales up the side so that the blue cap could be measured more accurately. Mining regulations stipulated that the following had to occur, based on the percentage of methane in the air:

- 1¼ per cent – all electrical equipment was turned off and shotfiring was prohibited
- 2 per cent – men were withdrawn to fresh air

The lamp had one further safety feature in that the flame would go out if oxygen levels reached 16 per cent, indicating the threat of suffocation. Such was the inherent accuracy of the safety lamp that mining engineers and colliery officials still carried the lamps as back-ups during the modern electrical era.

⬆ A display of miners' lamps, hanging from the ceiling of a public house. *(Author/Big Pit NCM)*

⬇ A modern mining electric lamp system, with all electrical elements sealed to ensure no external sparking. *(Author/Big Pit NCM)*

ELECTRICAL LAMPS

Electrical illumination, properly sealed and 'fire-safe' was the final step forwards in the evolution of safe mine lighting. The main challenge at first was to find adequate battery power while remaining conveniently manportable. The first portable electric lamps began to enter service in Britain and Europe in the 1880s (although an electric lamp had been approved for mine use back in 1859, but not adopted), but their illumination was poor and their battery packs unwieldy.

Electric lamps really took off in 1913, when the Oldham Type C safety lamp was given Home Office approval for mining use. By the following year, there were 75,700 electrical lamps of all types in British mines, powered by either lead-acid or alkaline batteries, and by 1934 that number had grown to 394,820.

⬆ **A lamp room. In recent times, mining illumination has increasingly shifted to Light Emitting Diodes (LED) bulbs with lithium batteries.** *(Author/Big Pit NCM)*

➡ **The Oldham electric cap lamp led the way in personal illumination for miners, introduced in the 1940s.** *(Author/Big Pit NCM)*

In addition, mines also started to receive fixed overhead electrical lighting.

The early generation of miners' lamps were lantern-style devices, hung from hooks or nails on the wall or located on the ground next to the miner. This meant that often the light did not fall directly where the miner was looking. Working in conditions of poor illumination could cause nystagmus – a condition in which repetitive and involuntary movements of the eye reduce vision and depth perception – so getting better lighting at the coalface was a priority health issue.

The big advance in mining lamp illumination was the invention of the battery cap lamp in the 1930s, an invention that began in the USA but which gradually spread to Britain. By hanging the battery pack on a waist belt and fitting the lamp on the front of the helmet, the direction of illumination was automatically guided by the direction of the wearer's gaze. The cap-mounted lamps were also significantly brighter than the lantern lamps, throwing out a more intense direct beam.

In Britain, the power for the head lamps was initially delivered by alkaline batteries that were prone to leaks, inflicting chemical burns if the leakage came into contact with the skin. Following nationalisation, however, all such batteries were replaced with lead-acid types. When the miner completed his shift, he would return his lamp kit to the lamp room, where the battery would be recharged on a rack so it would be ready for the next use.

GAS-DETECTION TECHNOLOGIES

In addition to the safety lamp, the second half of the 19th century and the early 20th century saw several other types of gas-detection equipment become available for use. *Practical Coal-Mining* of 1908 lists a handful of the devices available at this particular juncture of history, some more convincing than others. Notably, many of the examples come from French, German or Belgian mining engineers; the inventive interaction between Britain and the Continent was constant. The examples are mostly different species of lamp, more specifically configured to test for methane rather than provide illumination, using alcohol, hydrogen or benzine as fuels for the testing flame. Interestingly, there are also references to a significant number of electrical testing devices. One explained in some detail is the 'Living's Fire-damp Indicator':

> *'Two small loops of platinum wire of equal magnitude can be made to glow by passing an electrical current through them simultaneously. The current is generated by turning the handle of a small dynamo contained in the apparatus. One of the loops is surrounded by an air-tight brass cylinder, with a glass disc at one end; the other by a cylindrical copper gauze, also with a glass disc at one end. The two glass discs face each other at a distance of a few inches apart, and when the wires are made the glow of the light from each falls upon the two sloping surfaces of a small movable A-shaped block covered with white paper which stands between them.*

An electronic firedamp detector. Manuals recommend that testing is conducted at at least 10yd (9m) from the active coalface. *(Author/Big Pit NCM)*

> *When there is combustible gas in the air which circulates through the wire-gauze cylinder, the loop within that cylinder glows with greater intensity, or, in other words, emits more light than the opposite loop, in consequence of the combustion of the inflammable gas in contact with it. As a result, that side of the movable block nearest to it is brighter than the other side.'*
>
> **(Boulton 1908: 79)**

Boulton notes that the instrument was also fitted with a scale to measure the percentage of methane detected, but he commented that the action of turning the dynamo handle shook the generator sufficiently to make it difficult to tell which side of the block was actually the brightest.

As technology progressed into the 20th century, a new range of electronic and pump-action methanometers were introduced into pits. These devices made more precise analysis of the atmospheric conditions in the mine, and were produced in types that ranged from large case-mounted units through to pocket-sized instruments. In time, they would evolve into the highly sophisticated computerised gas analysers used in mines from about 1986 – instruments that within seconds could provide a complete and accurate breakdown of the air composition, with automatic alerting systems.

In balance to these devices, however, we should mention that during the 20th century a form of biological detector was also provided in the use of canary birds. (Prior to the introduction of canaries, mice were actually used for this purpose, although wild ones were preferred as tame mice typically just went to sleep in the carrier's hand. In the earlier shallow mines, it was also usual to lower a small dog down the shaft if the pit had been on stop for any length of time.) Although the use of canaries in coal mines for toxic gas detection seems at first glance an almost medieval practice, in reality it was a core safety procedure from 1911, when the birds were first introduced into British mines, until 1986, when the NCB officially announced that it was phasing out this practice. Until that time, two birds were typically kept at each mine specifically for this purpose. The idea of using the birds as what we term 'sentinel creatures' was first proposed in the mid-1880s by the Scottish physiologist J.S. Haldane. Haldane recognised that canaries had a high degree of sensitivity to variations in oxygen level, and thus they would quickly exhibit signs of distress, or even lapse into unconsciousness, if the air contained more than normal levels of methane and carbon dioxide. Thus, from 1911, canaries were carried down into the mines in cages, and were carefully observed by their handlers. If the birds suddenly looked weak or if they stopped singing, or if they dropped unconscious to the floor of the cage, the handler knew to begin evacuation procedures. The keeping of canaries was officially legislated by Mines and Quarries Regulations, and generally each mine would have two working canaries at any one time.

DUST MANAGEMENT

Dust management was a perennial issue in mining, and one that actually became worse during the age of mechanisation. While dust was an explosive threat, as already discussed, by far the greatest general danger it posed was inhalation diseases, which came to blight

This image, of miners operating a coal scoop, illustrates the filthy conditions even in a relatively modern mine. *(Author/Big Pit NCM)*

hundreds of thousands of miners' lives. By the 1950s, about 4,000 miners every year suffered pneumoconiosis, an industrial lung disease, and many went on to contract tuberculosis and other lung diseases, some of them fatal.

In terms of dust prevention, there were limits to what could be done, but there were some basic measures. All cutting tools had to be kept sharp and had to work efficiently, and compressed-air tools had to be monitored for dust-inducing air leaks. Blasting was kept to a minimum. Coal spillages had to be cleaned up quickly, and damaged tracks and roadways repaired to avoid spillages from passing cars and tubs. It was stipulated (if not always followed) that those same tubs and cars be filled to stated levels, as overfilling increased the frequency of spills. Interestingly, ventilation could actually exacerbate the dust problem, if the speed of air movement was too great. The *Deputy's Manual* recommended that the ideal air velocity at the face was 350ft (107m) per minute, and that this velocity should never exceed 500ft (152m) per minute.

In addition to dust prevention measures, there were a number of effective dust suppression processes and tools. The most important of these was the use of water to dampen the dust at the source of cutting. In 'wet cutting', water was sprayed directly on to the cutting jibs and wheels as the coal-cutting operation was underway, the water supplied either from a central feed or from a transportable pressurised water tank. As a general rule, about 2–5 gallons (9–23 litres) of water was required for each 1 yard (0.9m) of coalface worked.

A companion of wet cutting was 'wet drilling', in which water was sprayed around drills and powered picks when in use. Many of the machines came to have integral water sprayers, the water projected around the drill through side holes in the bit. An alternative to water was a foam suppressant, which had the added benefit of reducing the volume of water required. However, these were complex systems and were therefore not widely adopted.

Another technique of dust suppression was water infusion. Rather than projecting water on to the coal as it was being cut, this method worked by drilling boreholes into the coal seam, about 9ft (2.7m) apart and up to 7ft (2.1m) deep, and then pumping water into the holes from a high-pressure water infusion unit, driven by a hydraulic pump. With 15–30 gallons (68–136 litres) of water pumped into

To control dust emissions, water is sprayed from the head of this cutting drum as it works across the coalface. *(Author/Big Pit NCM)*

In West Virginia, United States, a coal miner sprays rock dust to reduce the combustible fraction of coal dust in the air. *(National Institute for Occupational Safety and Health)*

each hole, the coalface became internally soaked, thus when it was cut the volume of dust produced was significantly reduced.

In addition to the methods previously described, the liberal use of water sprays and mist projectors was a useful general tool of dust suppression. The spray units would ideally be used at all those points where dust was generated most intensively: the coalface, transfer points and loading stations on the haulage routes, and along conveyors.

MINING DISASTERS – COMBINED EFFECTS

Before we turn away from the topics of dust, gas, ventilation and explosions, a case study from mining history graphically illustrates how all of these issues were critically interrelated in the mechanics of a disaster. The great English Victorian writer Charles Dickens is best known for his novels, but as a reporter he also covered all manner of human-interest stories. In 1850, while writing for *Household Words*, a two-penny weekly magazine he had founded, Dickens recorded the words of several miners who had been involved in serious underground accidents in the collieries of north-east England. Here is an extract from one of those accounts, relating the experience of a miner at Willington Colliery, Durham, on 19 April 1841. At the time of the accident, the miner was working on the Bensham seam, and they were about 840ft (256m) below ground and 280 yards (256m) from the shaft bottom. Events took a rapid turn for the worse:

> *'A sudden rush of wind and dust came past us. It put out our candles. We knew directly there had been an explosion somewhere, and we ran along in the dark as fast as we could. We fell down several times, tumbling over stones and large pieces of coal or timber that had been shaken and blown out. When we got to the foot of the shaft, we found the iron cage stuck fast, all jammed*

with the explosion; but we made the signal, and another cage was lowered to us, into which we jumped, before it reached the bottom, by scrambling up the sides of the shaft. When we got to the bank, and had taken our breath a bit, we saw the chief viewer of the pit come running to us with his Davy lamp. We each took a Davy, and went down the pit, to see who we could help. We knew there had been sad work among them. When we got down to the bottom of the shaft, we soon heard moans and groans. They were two lads, still alive. We got them hoisted up in the cage to the bank; but they lived a very little while. Soon after, we found two more quite dead, shockingly burnt. We had not gone much further when we found there had been a great fall of the roofing; and among the loose coals and stones, and timbers we found a horse and a pony, all mangled and singed.'

Of note in this account is how the explosion, while obviously of a localised source somewhere in the network of tunnels, has a systemic effect on the mine, spreading its destructive force widely and blowing down 'coal and timber', even jamming the cage at the bottom of the shaft. (This exemplifies the reason why it became law for there to be two points of access to any mine.) The opening words of this account also point to the mechanism of coal-dust explosion that were outlined earlier. With the explosion over, however, the miners faced a new danger:

'We now met the after-damp, and were thinking of returning, when a groan made us go forward, and we brought out the body of a young man alive, but in such a state, be couldn't be recognised. We now found that the doors of the trappers in several places had been blown out, and consequently the air currents had ceased to ventilate all the west and north workings, so that those who were there, and had escaped the explosion, would be likely to lose their lives by the after-damp. A strange smell of burning now made us know that some other sort of fire was at work, and as we ran in the direction it smelt like burning straw, which told us it was the stables as had taken fire. And sure enough, there were all in thick yellow smoke and red flames. The horses were prancing wild about, and one, who was blind, got out, and tore away, and killed himself by running agen a wall. We all saw death before us, if we couldn't master this fire; because if it communicated with the workings in the west and north, where the bad gas was, there would be another blow-up worse than the first. Mr. Johnson, the viewer, acted like a man. We all gave our minds to the work, and succeeded in stopping out, with wood and wet clay plaster, the entrances to these workings. Fire engines were then got down, and we continued to pump at the stables, and at the walls of coal which had took fire on each side, and after we had drenched them with water for several hours, the fire was put out. It took thirteen hours and more to do this. The main currents of air were restored as usual, and we then

An abandoned pumping engine house at the Prestonpans Coal Mines near Edinburgh, Scotland, the beam arm of the engine projecting from the top of the structure. *(CI Photos/Shutterstock)*

➡ **Many mines had lamp relighting stations, but devices such as the one shown here were used for safe lamp relighting in any part of the colliery.** *(Author/Boulton)*

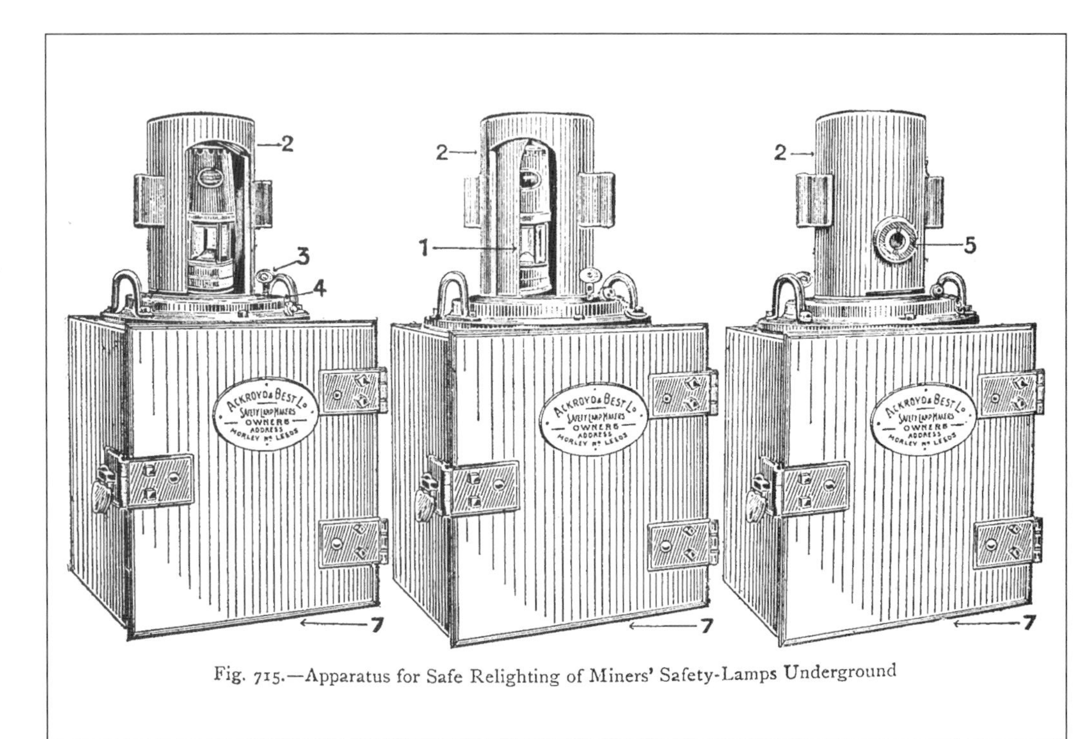

Fig. 715.—Apparatus for Safe Relighting of Miners' Safety-Lamps Underground

⬇ **Bodies are carried away at the site of the Senghenydd Colliery disaster, 14 October 1913. An explosion killed 439 miners.** *(Author/Big Pit NCM)*

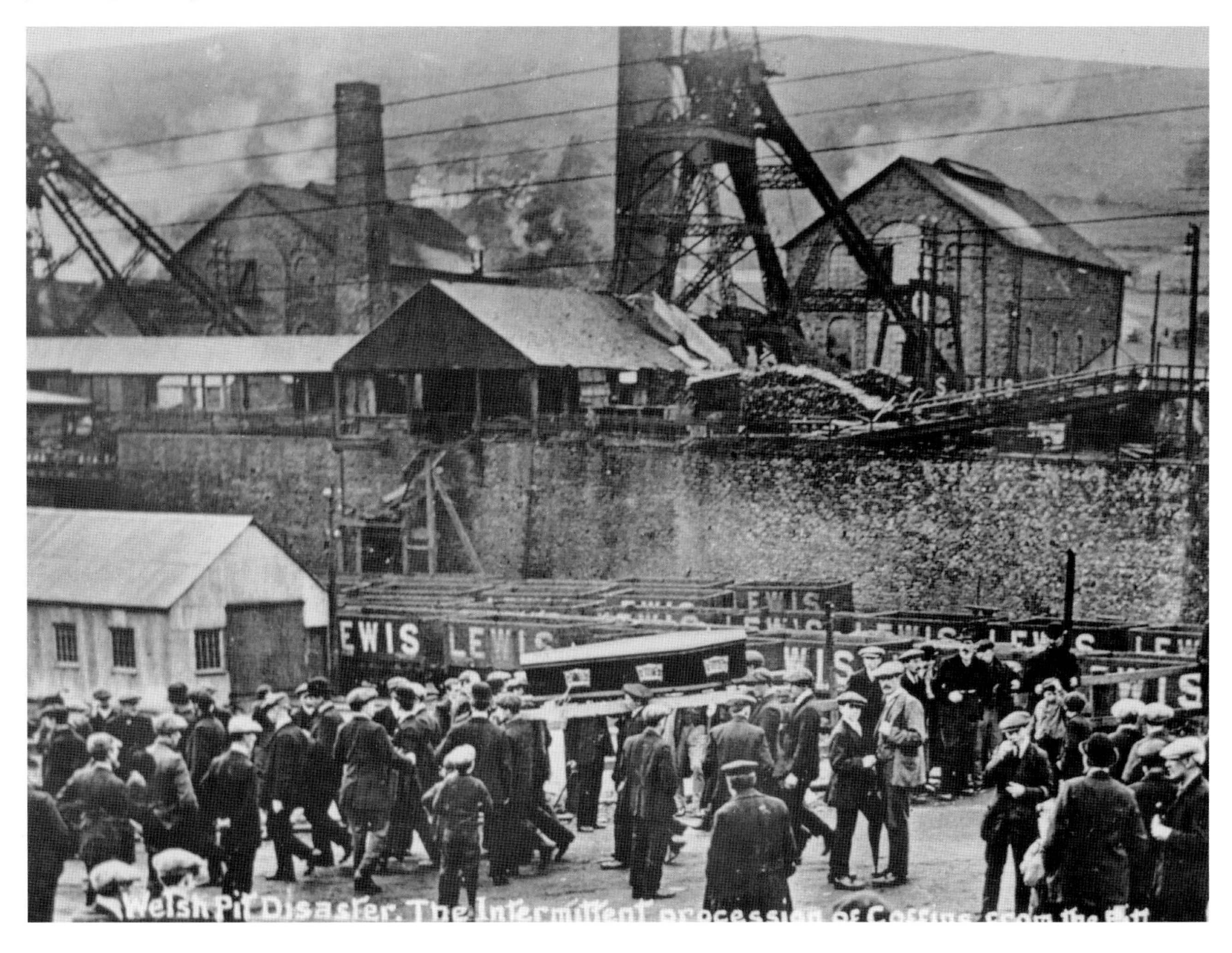

continued our search for those who had suffered by the explosion.'

This passage highlights both the human drama and the critical need for the miners to restore effective passage of ventilation. Explosions, by blowing open ventilation doors and regulators, changed the directional flow of air around the mine, which in this case both circulated afterdamp and threatened to blow methane-rich air on to the continuing fires, triggering further explosions. Working fast, the miners therefore improvised bashing-up walls with wood and plaster, modifying the airflow to assist the situation.

The physiological effects of afterdamp on the body were confronted in a different explosive disaster account recorded by Dickens, this time at Sloughton Colliery. The narrator of this event remembers the rapid-working effects of the poisonous gas after the explosion:

'As we ran we came upon others in the dark, and others came rushing out upon us from the side workings, and all of us together ran in a crowd and crush along the dark ways, in the direction of the shaft, and presently we found those who were foremost had fallen, and we got a sudden giddiness and gasping, so we knew we had met the choke-damp. It's a deathly, sleepy sickness you feel, and sinking at the knees, only you're sure it's not the breath of sleep you're afeeling, but you're breathing death. I called to those a-head to stop, and so did others near me, but many of them would go on, and down they went, one after the other.'

These accounts illustrate graphically both the dangers of the miner's profession, and also why generations of engineers pored over the issues of ventilation, dust and fire with such attention and innovation.

THE SELF-RESCUER

Introduced in 1967, by 1986 every miner was legally obliged to be given one of these instruments if working underground. The Self-Rescuer was a portable breathing-support device contained in a metal case and worn ready on the miner's belt; it weighed approximately 2.2lb (1kg). It was intended for use in conditions where the atmosphere contained lethal concentrations of carbon monoxide. In such circumstances, the miner would open the container, revealing the device and its mouthpiece; he would wear the Self-Rescuer around his head using the attached elasticated strap. Chemicals activated inside the apparatus converted carbon monoxide to breathable carbon dioxide, and could do so for a period of 90 minutes – just enough time for the miner to get to safety. It could not, however, provide protection in areas that were severely depleted of oxygen. One unsettling aspect of the apparatus was that the air breathed through it became very hot and dry, owing to the heat generated by the chemical reaction. The miner had to resist the temptation to remove the breathing kit temporarily for a gulp of cooler air, an action that could be fatal.

⬆ A Self-Rescuer device and its metal container, the container suspended in an accessible position from the miner's belt. *(Author/Big Pit NCM)*

⬆ **A length of exhaust ducting, used to extract dust and noxious gases, runs along the roof of a mine roadway.** *(Author/Big Pit NCM)*

➡ **The temporary props supporting a coal seam at the face could be of a rudimentary nature, here simple wooden props and braces.** *(Author/Big Pit NCM)*

ROOF SUPPORTS

Underground mining is essentially a fight against the geology of the planet. As soon as a shaft is sunk, or a roadway, gate or face is cut, the strata of earth and minerals around it begin their gargantuan effort to adjust to the sudden intrusion, and to fill the voids. This is not simply a matter of the roof attempting to sink downwards, although that is certainly the central concern, and the mathematics are worrying. For every 1ft (0.3m) of depth, the ground above exerts about 1lb per square inch (0.06 bar) of pressure. This means that at a distance of 1,500ft (457m) below ground – a depth commonly reached by many underground mines – the pressure is 1,500lb per square inch (103 bar), a truly formidable downforce. But at the same time as the roof is pressing downwards, the floor can be pressing upwards, the relative movement between the two surfaces being known as 'convergence'.

The problem is exacerbated by the way that the strata of rock above the void can, as it sinks downwards, start to express 'bed separation', with spaces opening up between the strata. That separation leads to even greater instability in the mass of rock above the void, structurally weakening the roof and leading to a possibly sudden and fatal collapse. Another unwelcome consequence of bed separation is that the spaces can fill up with firedamp, which suddenly explodes or is ignited during shotfiring operations. (Gas build-up behind the coalface in general can reach such high pressures that there are sudden violent eruptions from small points in the rock, propelling fragments of coal and stone at lethal velocities.)

For all the reasons described above, the practice of 'roof control' has historically been fundamental to coal mine operations. Even at relatively shallow depths, when the coal might appear self-supporting, the instability of the earth is such that any voids created need to be properly supported against collapse.

PACKS

The pack was one of the most substantial supports constructed in a mine, and it provided the main structural support to the underground workings. In its most basic form, a pack consisted of a dense, wide and deep section of stones, rammed into the space between the floor and the roof. A good pack was formed in two sections. First, an outer frame was built, consisting of 'walls' of large stones, laid so that their length pointed into

An NCB diagram from *The Support of the Roof at the Coalface* illustrates how chocks and supports are used to frame cutter tracks. *(NCB)*

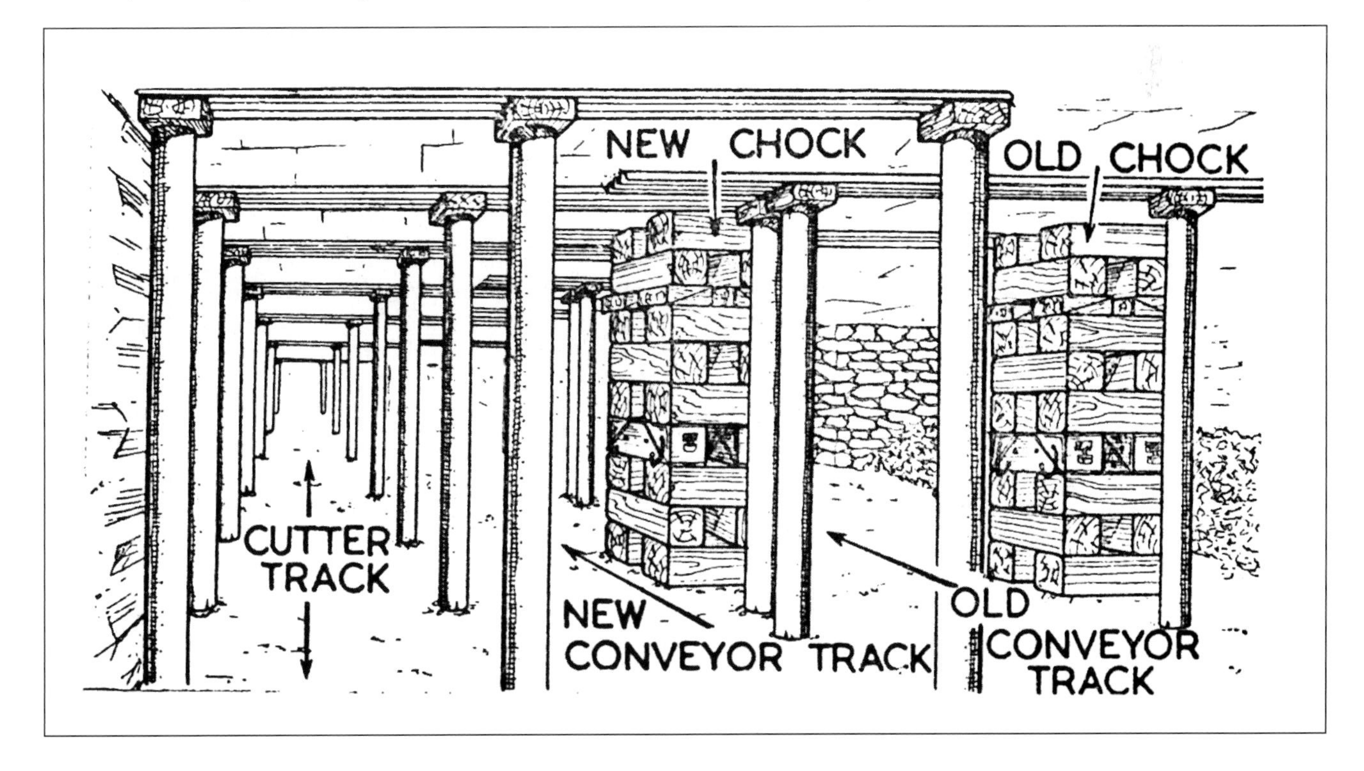

the interior of the pack. The space inside the walls was then filled with smaller stone debris. Other sections were built abutting the pack, using the same building process. Props and beams might frame and support the sides and corners of the pack; more modern packs might have outer walls of corrugated steel sheets, rather than stones.

The NCB manual *Roof Support and Control* (1961) explained that; 'The main support to the roof is provided by the packs which are constructed at intervals along the face and extend back into the waste [goaf]. They are also built on each side of the roadways. They support and control the roof in the waste and in the roadways like pillars supporting a bridge' (NCB 1961: 5).

The physical process of making packs could involve various different methods, from manual assembly by hand and shovel (rare in the age of mechanisation, and mainly confined to small packs) to the use of large chain-hauled scraper buckets. Specific methods of 'stowing' (i.e. using waste or other materials to support the roof) include:

- **Scraper packing** – This required the use of a scraper bucket, controlled via a winder and pulleys, collecting waste rock (known as 'ripping debris') and depositing it directly into the pack walls.
- **Hydraulic stowing** – This involved transmitting the packing material, in small pieces (sand, gravel etc.) via water pipes running down from the surface, pumping the excess water away once the pack had been created. Although this method was first used from about 1882, it did not gain widespread adoption, principally because of the damage the water did to surrounding floors and roofs.
- **Pneumatic stowing** – Dry stowing material was blown into the pack via pipes connected to a compressed-air blowing machine. The stowing material would usually have to be pre-crushed so

Trainee miners, under close tuition, learn the skills of installing wooden props and wedges. *(Author/Big Pit NCM)*

that it could feed successfully through the pipework, and the unit could be located either underground near the pack-making or on the surface.

- **Mechanical throwers** – These machines combined the crushing and filling functions; crushing hammers smashed up the stowing material, while scoop-like paddles collected this material and deposited it into the pack.

By the 1980s, pack-making received a technology boost in the form of pump-packing, which rapidly accelerated the speed with which packs could be created. The system's heart was a pump-packing station, which featured two mixing tanks and a hydraulic pump. The colour-coded mixing tanks contained two different cement-based powders (known as Techcem and Techbent), and these chemicals were mixed with water and then pumped out separately via 2.5in (6.4cm) pipes. At the opposite end of the pipes, where the pack was to be established, the pipework ran into a large canvas bag fitted with two inlet nozzles. The bag was first inflated with compressed air to allow the two chemicals to flow in and mix properly; a miner would make a cut in the top of the bag to allow the air inside to be squeezed out uniformly as the bag filled. The two chemicals, once inside the bag, began a chemical solidifying reaction, generating considerable heat as they did so. (Miners remember that you could almost burn your hand if you touched the outside of the bag while the hardening process was underway.) Within an hour or two, the packing materials had gone completely solid, forming a rapid pack. Although pump-packing was undoubtedly quick, it was also an expensive process, so economics meant that not every mine utilised this process.

CHOCKS

The purpose of a chock was to provide a solid support against the waste building up behind the working area of the coalface. The chock was formed from beams of hard wood (typically beech or oak), each measuring about 2ft (0.6m) long and 6in (15.2cm) square, built up in a right-angle lattice rather like a hollow 'Jenga' tower. At the top of the

⬆ A bank of hydraulic supports, *c.* 1970s. Note how the reach of the supports creates plenty of space for the shearer-loader. *(Author/Big Pit NCM)*

⬇ Alongside hydraulic supports, here we see a wooden chock, with the goaf pressed up behind. *(Author/Big Pit NCM)*.

Various forms of timbering used to support the roof and sides of a coalface and on roadways. *(Author/ Williams & Cryer)*

wooden stack were shallow wooden 'lids' (see below) to absorb some of the roof crush, and special steel release pieces could be built in approximately halfway up the stack; these beams featured angular corner sections that could be knocked out to release the pressure on the whole stack.

TIMBER AND STEEL PROPS

The oldest type of support was the wooden prop, which was effectively little more than a timber log cut to length, with either two square ends or a tapered end at ground level. Wood has its advantages for mining roof support. Despite its high strength, it also has

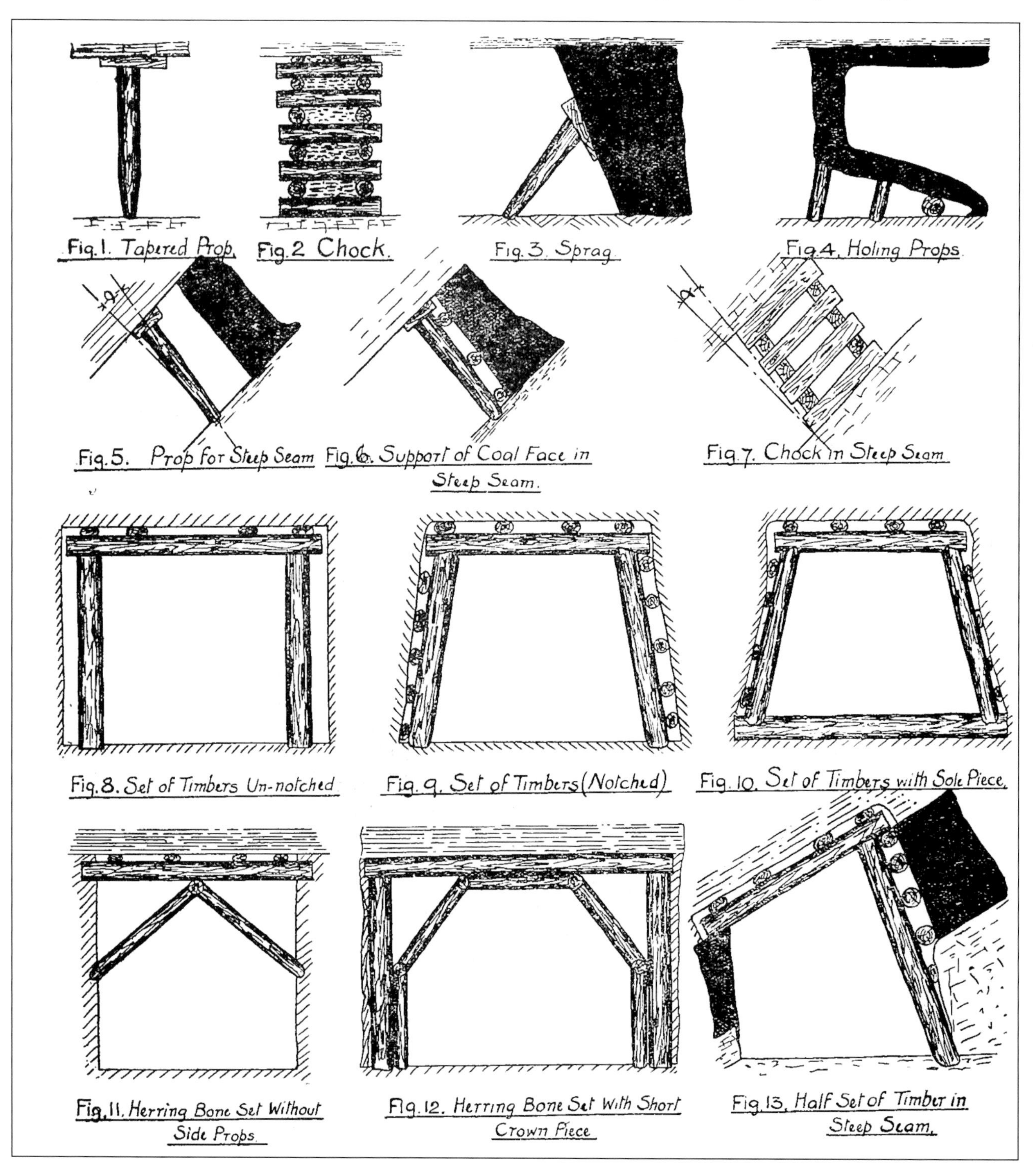

a natural degree of compressibility or 'yield', adjusting slightly to the pressure exerted upon it to develop a tight and accommodating fit between floor and roof. To fit a prop, however, a 'lid' – a small square section of soft wood – was inserted between the roof and the top of the prop, this lid both taking the crush of the roof while also helping to prevent the prop from slipping out.

A stronger and more modern alternative to the wooden prop was iron girders, of I-section for shorter props and H-section for longer props. In contrast to building girders, the mining girders typically had their flanges turned over at the top, so that the terminal edges of the girder did not cut into the roof when placed under pressure. An option instead of girders was tubular steel supports filled with concrete. As with wooden props, steel props still required a lid, wedge or other device to act as a pressure intermediary between prop and roof.

In contrast to the coalface, roadways were supported by prefabricated steel arches, which could support the roof while keeping the roadway completely open for haulage. In a typical modern mine tunnelling process, horizontal forepoling girders would be laid as temporary roof supports at the ripping lip (the advancing edge of the tunnel as it was being cut), and the arched supports were placed on top of these to form the permanent supports. The next set of forepoling girders was then hung off these arches to continue the process. An alternative to arched girders was to employ straight girders, set between robust stone or brick packs either side of the roadway.

Another important way of supporting both roofs and floors in coal mines was bolting. The principle of bolting was that a long steel bolt was driven, via a borehole, from a weaker outer strata up to stronger inner strata, stabilising

⬆ Timber is still used for roof supports in modern mines. Here two Ukrainian miners manufacture log props in situ in 2013. *(DmyTo/ Shutterstock)*

⬅ This miner is removing a support using a Sylvester prop withdrawer, the man obviously ready to move quickly backwards. *(Author/ Big Pit NCM)*

⬆ **The upright supports here are of the metal friction type, linked by articulated beams.** *(Author/Big Pit NCM)*

⬇ **Supports often followed the angle of the roof or downward pressure, rather than being strictly upright.** *(Author/Big Pit NCM)*

the rock. Although this process would not necessarily obviate the need for other supports, it could certainly improve the structural integrity of the heading. Various bolt heads were designed to improve the grip on the strata at the end of the bolt. Bolting, however, could induce fractures in the roof or floor if it was attempted in unsuitable strata, so was only performed after due geological investigation.

BARS

These were the support pieces laid flush with the roof between the individual upright supports, trapped between the lid and the roof, distributing the downward pressures over a wider area. They varied in size, but usually ranged from 4ft 6in (1.4m) to 7ft (2.1m) in length, and were either wooden or steel varieties. The wooden types were rectangular wooden planks or half-round timbers; the latter might be made by sawing wood props in half lengthwise.

Steel bars were often corrugated, to increase the overall thickness and weight resistance of the bar, but if the bar was required to be lengthy (as might occur on a prop-free front – see page 147) then steel girders of H-section or squared-off U-section might be applied.

HOLING NOGS, SPRAGS AND STAYS

Nogs or sprags refer to small supports that are used to prop up a coal seam as it was undercut. The classic holing nog/sprag was simply a wedge of wood, tapered at one end so that it could be hammered into place to support the weight of coal above. Steel varieties included those that replicated the overall shape of the wooden nog, but rendered in H-section steel girder, or L-shaped nogs whose upright bar also provided some support to the front face of the coal. Along a coalface it was not uncommon for there to be other sections of bulging or overhanging coal, and these were supported with lengthier sprags or stays, wooden beams or steel girders bracing the section of coal between its face and the floor.

HYDRAULIC SUPPORTS

We have seen many engineering revolutions throughout this book; regarding roof supports, the introduction of hydraulic props in 1947 was the key moment of transformation. Hydraulic props belong to the category of 'yielding supports', referring to the fact that they are not of rigid structure and that they were designed to 'give' if loaded beyond their maximum pressure, yielding to a point of stabilisation at about three-quarters of their maximum load.

The physics behind the hydraulic prop is the incompressibility of fluids. Each prop contained a given volume of fluid (typically a mix of 95 per cent water to 5 per cent oil), which the operator could place under pressure by cranking a lever that raised an internal piston. The NCB manual *Roof Support and Control* explains the science behind the system:

> *'The hydraulic prop makes it possible for a small force (provided by a man's arm) to exert by a pumping action a powerful thrust (by the piston action of the prop) against the mine roof. This is made possible because*
> *a) liquids are practically incompressible, and*
> *b) pressure applied anywhere on an enclosed liquid is transmitted instantly and undiminished throughout the liquid.*
>
> *The diagram [shown here] . . . shows how these principles enable a 10lb. weight to balance a 100lb. weight. The surface area of the liquid in the small vessel is 10 square inches and in the large vessel 100 square inches. The pressure in both vessels is therefore 1lb. per square inch (1 p.s.i.) and the weights are balanced. An increase in weight in the small vessel would force up the weight in the large vessel. The pressure*

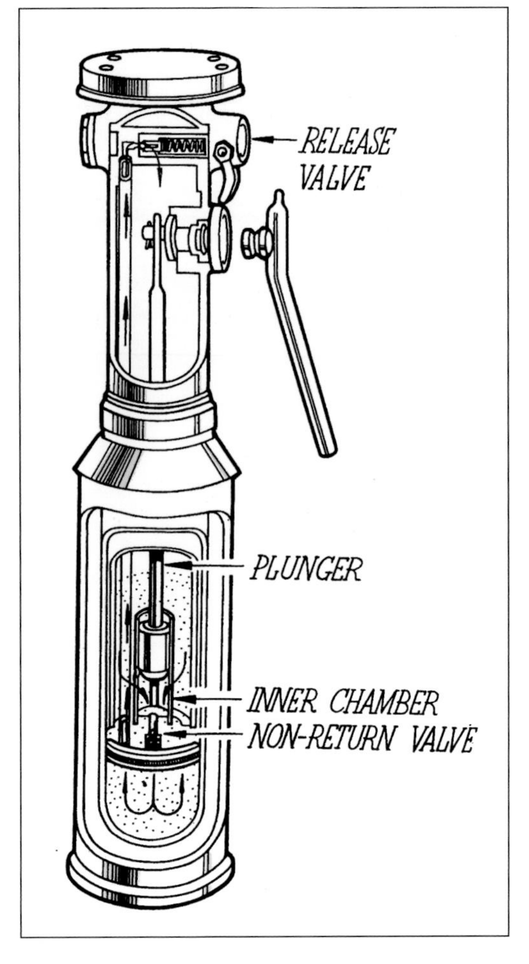

An internal diagram of a hand-operated hydraulic prop mechanism. *(NCB)*

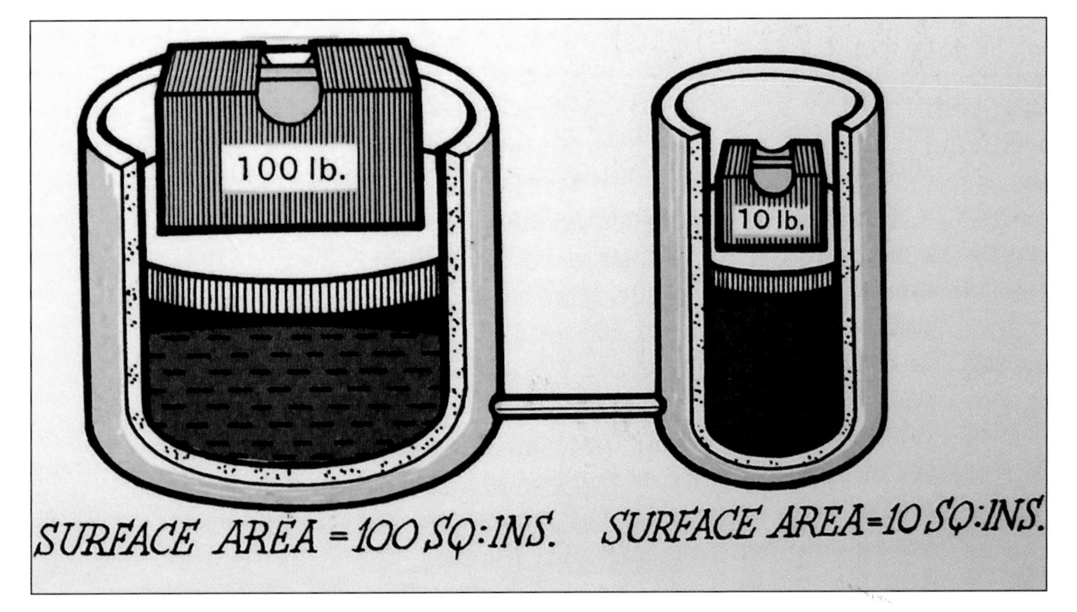

A diagram illustrating the scientific principle behind 'yielding' hydraulic supports. An increase in weight in the small vessel would raise the weight in the large vessel. *(NCB)*

➡ **W.E & F. Dobson (part of Hardwick Industries) began making hydraulic pit props in 1953.** *(Author/ Big Pit NCM)*

⬇ **A miner operates the pressure lever of a hydraulic prop to take up the strain against a steel girder above.** *(Cornwell/Big Pit NCM)*

of the operator's hand on the lever of the prop represents the weight in the small vessel pressing down, by means of a plunger, the liquid in the chamber (within the fluid reservoir) and so raising the level of the liquid in the "large vessel" in the base of the prop. The pressure acts against the bottom of the upper tube and forces it against the mine roof.'

(NCB 1961: 11–12)

Utilising this hydraulic principle, the manual goes on to explain that the prop would exert an 'immediate pressure' of 5 tons (5.08 tonnes), which in itself is more than a wooden or steel prop could exert, and would maintain its resistance until it reached 20 tons (20.3 tonnes) of pressure, as the roof weight settles on to it. At this point, a relief valve in the upper part of the prop released a small amount of fluid back into the inner tube, with the result that the prop yielded a fraction of an inch while still maintaining the 20-tonne maximum pressure.

From a practical point of view, hydraulic props changed the business of roof support. Not only could they be installed quickly and easily, simply by being put in place and the lever on the side being cranked up, but they also offered, as we have seen, greater load-bearing strength. They could also be removed easily, by a relief toggle being pulled, the whole prop collapsing in a controlled fashion over a few seconds; there was none of the hammering and winching involved with removing rigid supports. Furthermore, the cushioned weight bearing of the hydraulic prop reduced the likelihood of the prop fracturing the roof.

The downsides of the hydraulic props were mainly to do with cost and sophistication. They had to be handled properly, as damage to their component parts or outer tube would either prevent the prop being operated, or could compromise its strength.

FRICTION PROPS

An alternative to the hydraulic prop, although mainly prior to the 1970s, was the friction prop. These exerted their pressure via the friction between two members, set in either a telescoping or sliding arrangement. To install them, the prop was first extended to its required length, at which point it was tightened against the roof or bar by a setting wedge, claw or screw jack. A clamp was then applied to set the prop firmly in place.

According to some engineering manuals, friction props could be unreliable, prone to slipping or slackening under load, although this problem was reduced with proper installation procedures, or the use of some of the more sophisticated friction prop types. They were, however, quick to fit and easy to release, plus they were generally more robust than the hydraulic props, and could be repaired more easily.

THE PROP-FREE FACE

Any coal workings, whether longwall or pillar-and-stall, had to be supported rigorously and logically, with packs, chocks and props set at regular intervals around the coalface and working area. In traditional longwall working, the props were set in parallel rows between the seam on which the miners were working and the packs, chocks and goaf to their rear, the props essentially forming two colonnades or 'tracks': one for the cutting machine, one for the conveyor. The biggest issue with this arrangement was that the need to support the roof directly above the seam left little space for the cutting machine, usually no more than 3ft (1m) in depth, presenting obstacles that the cutting process had to negotiate. Also, the process of advancing the props after each cut was time-consuming.

The advent of hydraulic props offered the prospect of the 'prop-free face', meaning that the entire face was free of props from the coal seam to behind the conveyor belt. This was achieved by fitting a cantilever beam to hydraulic supports; when the supports were raised, the cantilever beam extended forwards to take the weight of the roof over the working area.

This principle came to be mechanically embodied in what are known as 'self-advancing supports', an impressive feature of the continuous mining operations introduced

As the inset here shows, emplacing supports in low seams was not a job for the claustrophobic. *(Author/ Big Pit NCM)*

from the 1950s. The individual self-advancing support varied in its construction between makes and models, but in overview it consisted of two to six hydraulic legs fitted to one or two long cantilever bars, the whole system fitted to a platform with an attached hydraulic jack, designed both to push forwards the conveyor system and to draw forwards the ram platform. Multiple such platforms were arranged in a continuous and unbroken row of supports across the coalface. Each support leg was designed to yield at a roof weight of between 22 and 33 tons (20 and 30 tonnes), thus the entire bank of supports provided an exceptionally solid overhead support for operations.

SELF-ADVANCING SUPPORT – OPERATION

The design and operation of the self-advancing support system was described in clear terms in the NCB manual *Roof Support and Control*:

> *'"Self-advancing" supports are used on faces where the coal is won by continuous mining methods. The supports do two jobs besides supporting the roof: They push over the sections of the flexible conveyor as the coal getting machine passes and draw themselves forward to support the newly exposed roof. Units of a self-advancing system consist of a number of hydraulic props linked together by a common hydraulic system supporting a strong cantilever beam which juts out over the conveyor to support the roof right up to the coalface. They are attached to the flexible conveyor by a hydraulic jack which is used both to push over the conveyor and, afterwards, to draw forward the support. The operation, governed by a set of hydraulic controls, follows this sequence: 1. Push forward the conveyor by the hydraulic jack when the coal getting machine has passed; 2. Lower roof beam; 3. Pull the support forward by the jack; 4. Raise the beam to support the roof in the new position. In practice only one support in three, four or five may be used to push over the conveyor. The supports are usually advanced alternately to maintain even support for the roof and maximum protection for the men. On many faces the roof behind the self-advancing supports is allowed to collapse freely into the waste. This is called "caving".'*
>
> **(NCB 1961: 14)**

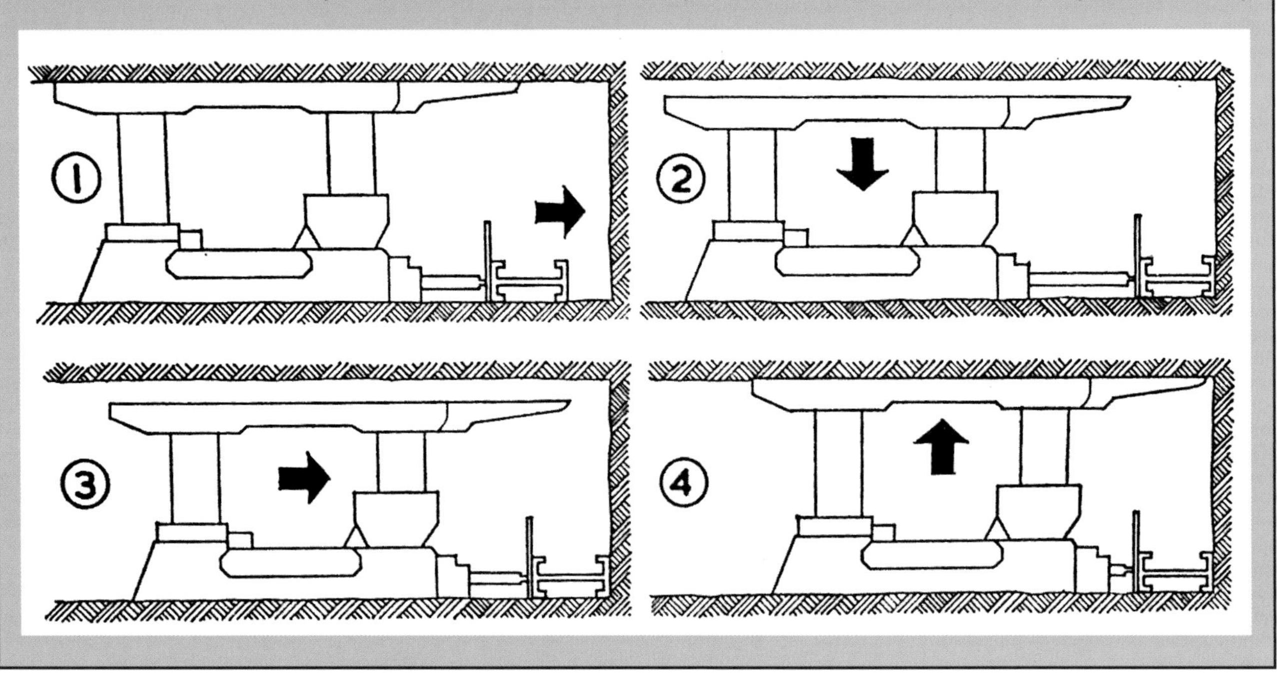

An NCB diagram showing the principle of the self-advancing support, the hydraulic cantilever beams inching forward as the face is cut. *(NCB)*

The true benefits of the self-advancing supports, however, were the open working space on the coalface for both cutter/shearer and conveyor and the way that the props could advance forwards with the workings without manually setting a multitude of supports. It was another aspect of mechanisation that took the British coal mines towards maximum possible output.

DRAINAGE

Drainage was another of those areas of mining science of arcane but crucial importance. The ground beneath the surface was rarely composed of pure strata of unbroken and dry rock. Surface inundations of rainwater

⬆ **A close-up of modern powered supports, the hydraulic leg bracing upwards to the cantilever beam. Note also the conveyor ram connecting to the base of the shearer-loader frame.** *(Author/Big Pit NCM)*

⬅ **From a 1797 edition of *Encyclopedia Britannica*, a diagram of a Watt steam engine, which had ready applications in mine drainage.** *(DigbyDalton/ CC BY-SA 3.0*

➡ At the Beamish Museum, County Durham, England, a selector lever for changing the speed of the winding engine between men and coal. In the post-coal age in Britain, the image carries some symbolic weight. *(David P Howard / Men or Coal? / CC BY-SA 2.0)*

would usually soak into the terrain below, although generally speaking the deeper the mine, the less this became problematic. The subterranean strata might therefore be riven through with water-filled faults, fissures and natural reservoirs, sources of water that would be unleashed when opened by the mining process. At best, the water might be a constant trickle from the coalface, easily managed by basic pumping measures. At worst, the water could become a torrent that threatened both the mining operations and the miners.

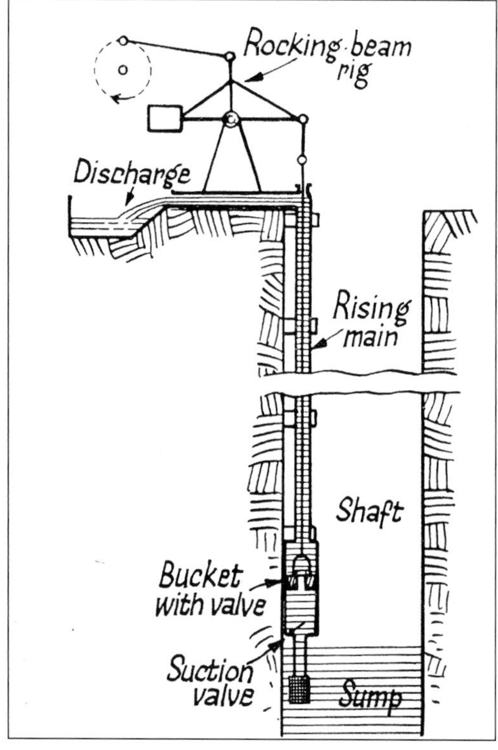

➡ A diagram of a simple shaft pump, which uses a piston and atmospheric pressure to draw the water to the surface. *(Author/Mason)*

The earliest form of mine drainage technique was natural drainage, cutting tunnels or adits below the level of the main workings, and sloping them downwards so the water ran freely down the slope until it exited, typically from the side of a hill or the base of a valley. The location of the mine and the geology of the workings were often not amenable to this drainage, however, so other measures were taken. In a water-logged mine, a water-collection sump might be dug below the mine shaft; the water from the workings would drain into the sump, from where it would be transported to the surface in buckets. During the 16th century, this system was given an efficiency boost in the 'chain and rag' mechanism. Leather balls or discs were cycled on an endless chain through a pipe that ran through the sump; the water was trapped in the pipe above the balls, and was lifted to the surface. These might be powered by horse

⬅ **Water intrusion was, and remains, a constant problem to manage in mining. Precautionary boring is often conducted to detect large measures of water.** *(Author/Big Pit NCM)*

gins or waterwheels, which were also applied to other early pumping mechanisms.

As we explored earlier in this book, steam power changed the nature of mining on many levels, and drainage was no exception. Indeed, although steam came to be applied to winding, it was its applications for drainage that were of primary significance. The first installation of a Newcomen steam-pumping engine, at a colliery near Dudley in the West Midlands, was in 1712. These machines became relatively common, although only in mines with the deeper financial pockets to afford them. Later in the 18th century, and during the following century, improved Boulton and Watt steam pumps were also working hard in Britain's mines. One such pump, installed at Landore, Swansea, in 1800, had a 25hp engine with a 78in (1.98m)-diameter cylinder, a piston stroke of 8–9ft (2.4–2.7m) and a capacity to move 1,100 gallons (5,000 litres) every minute with every 12–13 strokes (RCAHMW 1994: 87).

Pumping technology underwent a major improvement in the late 18th and early 19th centuries with the introduction of hydraulic ram principles, in which the water was forced, rather than lifted, to the surface, meaning that the weight of the steam beam arms contributed to the efficiency of the pumping (RCAHMW 1994: 88).

Steam-pumping technology thereafter went through various stages of evolutionary improvement, including the adoption of the high-pressure Cornish boiler, the centrifugal pump and rotative engines. By the end of the century, however, electric pumps were starting to come into service. The efficiency of these systems meant that there were few underground mines where drainage was a major problem, at least in the sense of stopping production.

Drainage technology is a subject of great complexity, with numerous different types, makes and models of pumping engines. This caveat could be applied to almost any topic that we have covered in this book. Although the underground coal mining industry in Britain is, at least at the time of writing, no more, the technological innovation and the human activity that made it possible should forever be admired. Without the mines, without the coal, Britain as a nation would have achieved but a fraction of what it eventually accomplished.

AFTERWORD

For more than two centuries, Britain was carried aloft on the muscular, pitch-black shoulders of 'King Coal'. Without coal, we would not have had an Industrial Revolution, an empire, a steel industry, a mighty navy, national electrical power – were it not for coal, British history would be unrecognisable from what it actually became. This dirty mineral powered the nation, and indeed the world.

But coal mining in Britain is more than just a grand narrative of national and industrial might. The true heart of coal's story is its people, millions of them, both the miners who toiled and the families they supported and who, in turn, supported them. An individual colliery was not just a locus of work, but the epicentre of a far larger human experience. Pits bound together what is today that most elusive of entities – communities. These communities were self-contained worlds, pinned in place by the headgear around which their lives revolved. More than just employment, the mines generated cultures, each community having its own history, outlook, rhythms, housing, tragedies, art, music, language. They also fostered identity, pride and purpose, qualities that would in time be sorely lost.

My own grandfather, Ernest Davison, went down the mines of South Yorkshire when he was 13 and spent his entire working life in the industry, moving from underground worker to surface fitter. The work certainly hardened him. Boys very quickly became men in those days. Men in turn continued the accelerated ageing through long hours in a dangerous and dust-choked subterranean world, with the physical pressures of a planet literally attempting to slam shut the thin gap in which they worked. My grandfather was a quiet man, his eyes often lost in the embers of the hearth, which burned the very coal he helped to extract. Occasionally, however, he would share stories of the miner's life, and they both enthralled and appalled me. Huge tree-trunk pit props reduced to splinters by the geological compression. Never seeing daylight for weeks on end during the winter. Attending ghastly accidents at the bottom of yawning pit shafts. My grandfather never romanticised mining work, but around this often-lethal industry, he – along with my equally redoubtable grandmother Dorothy – built a home and a family, as did hundreds of others on the mile-long street on which they lived.

Coal mining still pulses in many parts of the world, albeit often under conditions of industrial decline. Advanced computerised technologies have revolutionised safety, scale and productivity, at least in the most advanced and well-funded mines. In the UK, however, all the deep mines all silent; the last closed in 2015. Furthermore, the country is driving away from the legacy of coal with increasing speed, the motion spurred not only by legitimate concerns regarding the environmental impact of coal as a fuel, but also by advances in new power sources. On 21 April 2017, Britain hit a landmark with the first 24-hour period since the Industrial Revolution in which all Britain's electrical power was provided without burning coal. In 2019, a record of 18 days of coal-free powered generation was set, and under current government plans, it is likely that all remaining coal-fired power plants in the UK will be closed by 2025. King Coal will almost certainly never wear its crown again.

While many excellent national and local museums keep the story of mining alive, the landscape of Britain is also dotted with dozens of headframes, the sheave wheels that once thrummed day and night now rusted into silence and immobility. This book has focused primarily on the technological and engineering dimension of coal mining, but the social and culture history is a vast, rich topic in its own right, one deserving of our reading and research. Our modern nation was built on coal and the men who won it. So, when you next pass a headframe, pause for a moment to give it your full attention, and think of those who kept it, and an entire nation, turning.

Chris McNab

⬅ Miners in a cage, their clean faces indicating the start of the shift, ready for their descent to pit bottom. *(Author/Cornwell/Big Pit NCM)*

FURTHER READING

Ashworth, William and Mark Pegg (1986). *The History of the British Coal Industry. Vol.5: 1946–82 – The Nationalized Industry. Oxford:* Oxford University Press.

Boulton, W.S (1908). *Practical Coal-Mining. Vols 1–6.* London: Gresham Publishing Company.

Ezra, David (et al.) (1976). *Coal: Technology for Britain's Future.* London: Macmillan.

Farey, John (1811). *General View of the Agriculture and Minerals of Derbyshire; With Observations on the Means of their Improvement.* London: Board of Agriculture.

Greenwell, G.C (1888). *A Glossary of Terms used in the Coal Trade of Northumberland and Durham.* 3rd edition. London: Bemrose and Sons.

Hayes, Geoffrey. *Coal Mining.* Oxford: Shire Publications.

Hayman, Richard (2016). *Coal Mining in Britain.* Oxford: Shire Publications.

J.C. (1708). *The Compleat Collier: Or, The whole Art of Sinking, Getting, and Working, Coal-Mines, &c. As is now used in the Northern Parts, Especially about Sunderland and New-castle.* London: G. Conyers.

Mason, E. (1956). *Deputy's Manual.* Vols 1–2. London: Virtue and Company Ltd.

Nature, (1938). 'Electricity in Coal Mines.' 142/989. https://doi.org/10.1038/142989b0

NCB (1951). *The Installation and Extension of a Gate Belt Conveyor.* Training Manual No. 3. London: National Coal Board.

NCB (1952a). *Handbook on Shotfiring in Coal Mines.* London: National Coal Board.

NCB (1952b). *Pit Pony.* London: National Coal Board.

NCB (1952c). *The Operation of a Longwall Cutting Machine.* Training Manual No. 4. London: National Coal Board. Accessed at http://www.dmm.org.uk/books/trn04-00.htm

NCB (1957). *Moving Coal Underground.* London: National Coal Board.

NCB (1961). *Roof support and control.* London: National Coal Board.

NCB (1966). *Powered Supports: a mechanisation training booklet.* London: National Coal Board.

NCB (1960). *Colliery Ventilation Officer's Handbook.* London: National Coal Board.

NCB (1972). *The Support of the Roof at the Coalface.* London: National Coal Board Industrial Training Branch.

NCB (1976). *Transport Underground: Part 1 – Moving out the coal.* London: National Coal Board.

NCB (1976). *Transport Underground: Part 2 – Moving men and supplies.* London: National Coal Board.

NCB (1977). *Armoured Flexible Conveyors.* London: National Coal Board Industrial Training Branch.

NCB (1950). *Pit Ventilation.* London: National Coal Board.

Preece, Geoff (1981). *Coalmining. Salford: Salford Museum of Mining.*

RCAHMW (1994). *Collieries of Wales: Engineering and Architecture: Aberystwyth: Royal Commission on the Ancient and Historical Monuments of Wales.*

RCHM (1994). *Images of Industry: Coal.* Swindon: Royal Commission on the Historical Monuments of England.

Singh, R.D. (2019). *Principles and Practices of Modern Coal Mining.* New Delhi: New Age International Publishers.

Tomalin, Miles (1960). *Coal Mines and Miners.* London: Methuen.

Thompson, Ceri (2008). *Harnessed: Colliery Horses in Wales.* Cardiff: National Museums Wales Books.

Williams, M.D. and Arthur Cryer (1918). *Mining Sketches.* Cardiff: Williams and Cryer.

INDEX

Page numbers in *italics* indicate illustration captions.